A Guide to Real Variables

Library of Congress Catalog Card Number 2009929076
ISBN 978-0-88385-344-3
Printed in the United States of America
Current Printing (last digit):
10 9 8 7 6 5 4 3 2 1

The Dolciani Mathematical Expositions
NUMBER THIRTY-EIGHT

MAA Guides # 3

A Guide to Real Variables

Steven G. Krantz
Washington University, St. Louis

Published and Distributed by
The Mathematical Association of America

The DOLCIANI MATHEMATICAL EXPOSITIONS series of the Mathematical Association of America was established through a generous gift to the Association from Mary P. Dolciani, Professor of Mathematics at Hunter College of the City University of New York. In making the gift, Professor Dolciani, herself an exceptionally talented and successful expositor of mathematics, had the purpose of furthering the ideal of excellence in mathematical exposition.

The Association, for its part, was delighted to accept the gracious gesture initiating the revolving fund for this series from one who has served the Association with distinction, both as a member of the Committee on Publications and as a member of the Board of Governors. It was with genuine pleasure that the Board chose to name the series in her honor.

The books in the series are selected for their lucid expository style and stimulating mathematical content. Typically, they contain an ample supply of exercises, many with accompanying solutions. They are intended to be sufficiently elementary for the undergraduate and even the mathematically inclined high-school student to understand and enjoy, but also to be interesting and sometimes challenging to the more advanced mathematician.

1. *Mathematical Gems,* Ross Honsberger
2. *Mathematical Gems II,* Ross Honsberger
3. *Mathematical Morsels,* Ross Honsberger
4. *Mathematical Plums,* Ross Honsberger (ed.)
5. *Great Moments in Mathematics (Before 1650),* Howard Eves
6. *Maxima and Minima without Calculus,* Ivan Niven
7. *Great Moments in Mathematics (After 1650),* Howard Eves
8. *Map Coloring, Polyhedra, and the Four-Color Problem,* David Barnette
9. *Mathematical Gems III,* Ross Honsberger
10. *More Mathematical Morsels,* Ross Honsberger
11. *Old and New Unsolved Problems in Plane Geometry and Number Theory,* Victor Klee and Stan Wagon
12. *Problems for Mathematicians, Young and Old,* Paul R. Halmos
13. *Excursions in Calculus: An Interplay of the Continuous and the Discrete,* Robert M. Young
14. *The Wohascum County Problem Book,* George T. Gilbert, Mark Krusemeyer, and Loren C. Larson
15. *Lion Hunting and Other Mathematical Pursuits: A Collection of Mathematics, Verse, and Stories by Ralph P. Boas, Jr.,* edited by Gerald L. Alexanderson and Dale H. Mugler
16. *Linear Algebra Problem Book,* Paul R. Halmos
17. *From Erdős to Kiev: Problems of Olympiad Caliber,* Ross Honsberger

18. *Which Way Did the Bicycle Go? ... and Other Intriguing Mathematical Mysteries,* Joseph D. E. Konhauser, Dan Velleman, and Stan Wagon
19. *In Pólya's Footsteps: Miscellaneous Problems and Essays,* Ross Honsberger
20. *Diophantus and Diophantine Equations,* I. G. Bashmakova (Updated by Joseph Silverman and translated by Abe Shenitzer)
21. *Logic as Algebra,* Paul Halmos and Steven Givant
22. *Euler: The Master of Us All,* William Dunham
23. *The Beginnings and Evolution of Algebra,* I. G. Bashmakova and G. S. Smirnova (Translated by Abe Shenitzer)
24. *Mathematical Chestnuts from Around the World,* Ross Honsberger
25. *Counting on Frameworks: Mathematics to Aid the Design of Rigid Structures,* Jack E. Graver
26. *Mathematical Diamonds,* Ross Honsberger
27. *Proofs that Really Count: The Art of Combinatorial Proof,* Arthur T. Benjamin and Jennifer J. Quinn
28. *Mathematical Delights,* Ross Honsberger
29. *Conics,* Keith Kendig
30. *Hesiod's Anvil: falling and spinning through heaven and earth,* Andrew J. Simoson
31. *A Garden of Integrals,* Frank E. Burk
32. *A Guide to Complex Variables* (MAA Guides #1), Steven G. Krantz
33. *Sink or Float? Thought Problems in Math and Physics*, Keith Kendig
34. *Biscuits of Number Theory*, Arthur T. Benjamin and Ezra Brown
35. *Uncommon Mathematical Excursions: Polynomia and Related Realms*, Dan Kalman
36. *When Less is More: Visualizing Basic Inequalities*, Claudi Alsina and Roger B. Nelsen
37. *A Guide to Advanced Real Analysis* (MAA Guides #2), Gerald B. Folland
38. *A Guide to Real Variables* (MAA Guides #3), Steven G. Krantz

MAA Service Center
P.O. Box 91112
Washington, DC 20090-1112
1-800-331-1MAA FAX: 1-301-206-9789

To

G. H. Hardy and J. E. Littlewood,

our role models

Contents

Preface

Most of ancient and medieval mathematics concerned geometry and algebra. Questions of analysis arose in the work of Euler and Stirling and others, but only in isolated morsels. The real need for analysis became apparent when Newton and Leibniz's calculus took hold. This powerful new set of tools required some theoretical underpinning, some rigorous foundation, and analysis was the tool that was needed to carry out the program.

It was not until the nineteenth century that the necessary talent and focus came together to produce analysis as we know it today. Cauchy, Weierstrass, Riemann, and many others laid the foundations of the subject, provided the necessary definitions, and proved the required theorems. In the twentieth century Zygmund, Besicovitch, Hardy, Littlewood, and many others have carried the torch and continued to develop the subject.

The importance and centrality of real analysis is certainly confirmed by the fact that virtually every graduate program in the country—indeed, in the world—requires its students to take a qualifying exam in the subject. We are exposed to real analysis, both at the undergraduate and graduate levels. Today, analysis has assumed a newly prominent position in the infrastructure because of many new engineering applications such as wavelets, and also new financial applications such as the Black-Scholes theory of option pricing.

The fact remains that real analysis continues to be a rather technical and recondite subject. This is in part because mastery of the discipline is more a matter of technique than erudition or conceptual development. Generally speaking, the real analysis qualifying exam at any university is the hardest of all the quals. That is no accident, because problems in real analysis are tricky and demanding by their very nature. There is no royal road to real analysis.

The purpose of the present book is to provide an aid and conceptual support for the student studying for the qualifying exam in real analysis. This is a two-pronged effort. For this little volume concentrates on topics from undergraduate analysis; a separate volume, by another author, will treat topics from graduate analysis. More specifically, we discuss here sequences, series, modes of convergence, the derivative, the integral, and metric spaces. The graduate volume will treat measure theory, functional analysis, Fourier analysis, probability, and other advanced topics. The two volumes together will provide a unique and particularly friendly companion to the learning process in this subject area.

This book concentrates on concepts, results, examples, and illustrative figures. We downplay proofs, not because they are not important (they are, in fact, the essence of the subject), but because we want this to be a book that is easy to dip into and easy to take ideas from. The reader will use this text alongside a more traditional tome that provides all the dirty details. Our book is an entree to the subject area.

It is always a pleasure to write a book for the Mathematical Association of America, and to work with Editors Don Albers and Underwood Dudley. Technical Editor Beverly Ruedi is a master of her art, and makes the books sparkle with artistry. We also thank the MAA for engaging a team of particularly insightful and industrious reviewers who helped to keep the book on point and focused.

St. Louis, Missouri Steven G. Krantz

CHAPTER 1

BASICS

1.1 SETS

Set theory is the bedrock of all of modern mathematics. A *set* is a collection of objects. We usually denote a set by an upper case roman letter. If S is a set and s is one of the objects in that set then we say that s is an *element of* S and we write $s \in S$. If t is not an element of S then we write $t \notin S$.

Some of the sets that we study will be specified just by listing their elements: $S = \{2, 4, 6, 8\}$. More often we shall use *set-builder notation*: $S = \{x \in \mathbb{R} : 4 < x^2 + 3 < 9\}$. This last is read "the set of x in the reals such that $x^2 + 3$ lies between 4 and 9."

The collection of all objects not in the set S is called the *complement* of S and is denoted by ${}^cS$. The complement of S must be understood in the context of some "universal set"—see Example 1.1.

If S and T are sets and if each element of S is also an element of T then we say that S *is a subset of* T and we write $S \subseteq T$. If S is not a subset of T then we write $S \not\subseteq T$.

EXAMPLE 1.1.1. Let

$$S = \{a, b, c, d, e\}, \quad T = \{a, c, e, g, i\}, \quad \text{and} \quad U = \{c, d\}.$$

Then

$$a \in S, \ a \in T, \ d \in S, \ d \notin T, \ U \subseteq S, \ U \not\subseteq T.$$

If the universe is understood to be the standard 26-letter roman alphabet, then it follows that

$${}^cT = \{b, d, f, h, j, k, l, m, n, o, p, q, r, s, t, u, v, w, x, y, z\}. \qquad \square$$

1.2 OPERATIONS ON SETS

If S and T are sets then we let $S \cap T$ denote the collection of all objects that are both in S and in T. We call $S \cap T$ the *intersection* of S and T.

In case $S_1, S_2, S_3, \ldots$ are sets then the collection of all objects common to all the S_j, called the *intersection* of the S_j, is denoted by

$$\bigcap_{j=1}^{\infty} S_j \qquad \text{or} \qquad \bigcap_j S_j .$$

If S and T are sets then we let $S \cup T$ denote the collection of all objects that are either in S or in T or both. We call $S \cup T$ the *union* of S and T.

In case $S_1, S_2, S_3, \ldots$ are sets then the collection of all objects that lie in at least one of the S_j, called the *union* of the S_j, is denoted by

$$\bigcup_{j=1}^{\infty} S_j \qquad \text{or} \qquad \bigcup_j S_j .$$

Figure 1.1 illustrates the concepts of intersection and union, by way of what is known as a *Venn diagram.*

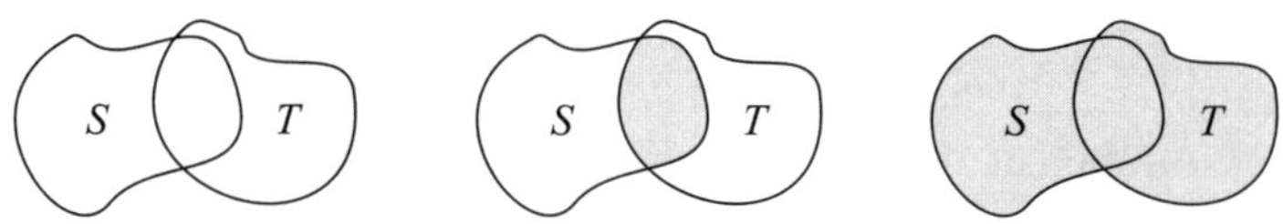

two sets S and T intersection union

FIGURE 1.1. Venn diagram of an intersection and a union.

EXAMPLE 1.2.1. Let $S = \{1, 2, 3, 4, 5\}$ and $T = \{2, 4, 6, 8, 10\}$. Then

$$S \cap T = \{2, 4\} \qquad \text{and} \qquad S \cup T = \{1, 2, 3, 4, 5, 6, 8, 10\} . \qquad \square$$

If S and T are sets then we let

$$S \times T \equiv \{(s, t) : s \in S \text{ and } t \in T\} .$$

We call $S \times T$ the *cartesian product* of S and T. Observe that $S \times T$ and $T \times S$ are distinct. Sometimes we will take the product of finitely many sets $S_1, S_2, \ldots, S_k$. Thus

$$S_1 \times S_2 \times \cdots \times S_k = \{(s_1, s_2, \ldots, s_k) : s_j \in S_j \text{ for all } j = 1, \ldots, k\} .$$

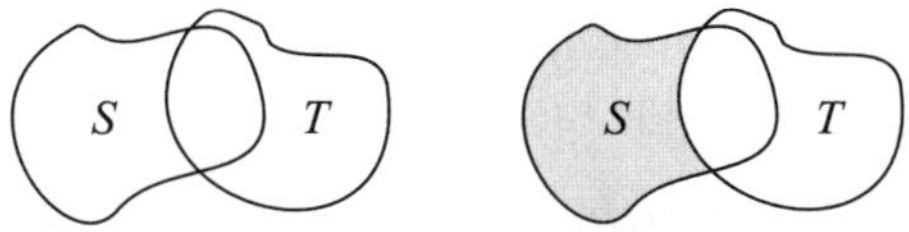

FIGURE 1.2. Venn diagram of a set-theoretic difference.

If S and T are sets then we let their *set-theoretic difference* be

$$S \setminus T \equiv \{x \in S : x \notin T\}.$$

If $S, T \subseteq \mathbb{R}$, the real numbers, then ${}^c S = \mathbb{R} \setminus S$ and $S \setminus T = S \cap {}^c T$. Figure 1.2 illustrates the concept of set-theoretic difference.

EXAMPLE 1.2.2. Let $S = \{a, b, 1, 2\}, T = \{b, c, d, 2, 5\}$, and $U = \{\alpha, \beta\}$. Then

$$S \setminus T = \{a, 1\} \qquad \text{and} \qquad T \setminus S = \{c, d, 5\}.$$

Also

$$S \times U = \{(a, \alpha), (b, \alpha), (1, \alpha), (2, \alpha), (a, \beta), (b, \beta), (1, \beta), (2, \beta)\}$$

and

$$U \times S = \{(\alpha, a), (\alpha, b), (\alpha, 1), (\alpha, 2), (\beta, a), (\beta, b), (\beta, 1), (\beta, 2)\}. \quad \square$$

We conclude by noting that there is a distinguished set that will arise frequently in our work. That is the *empty set* $\emptyset$. The empty set is the set with no elements. Observe that $\emptyset \subseteq A$ for any set A.

1.3 FUNCTIONS

Let S and T be sets. A *function* f from S to T is a rule that assigns to each element of S a unique element of T.

EXAMPLE 1.3.1. Let $S = \{1, 2, 3\}$ and $T = \{a, b\}$. The rule

$$\begin{aligned} 1 &\longrightarrow a \\ 2 &\longrightarrow a \\ 3 &\longrightarrow b \end{aligned}$$

is a function, because it assigns a unique element of T to each element of S. It assigns the same element of T to each of 1 and 2 in S; that is allowed.

$\square$

We write $f : S \to T$ if f is a function from S to T. We call S the *domain* of f and we call T the *range* of f.

EXAMPLE 1.3.2. Let $S = \{a, b, x\}$ and $T = \{1, \alpha, \gamma\}$. Define the function f by

$$f : \begin{cases} a \to \alpha \\ b \to 1 \\ x \to \alpha \end{cases}$$

Then the domain of f is the set S itself. There are several choices for the range. The set $\{\alpha, 1\}$ can be said to be the range. Also the entire set T can be said to be the range. □

Many of our functions are given by formulas. If we write, for example, $f(x) = \sqrt{2x+5}$, then we mean that the function f assigns to each number x the number obtained by doubling x and adding 5 and then taking the square root. We understand the domain of f to be all numbers x for which the formula defining f makes sense—for this example, the domain is $\{x : x \geq -5/2\}$. We understand the range of f to be any set containing all the values of f—for this example, the range could be taken to be $\{y : y \geq 0\}$.

If the function f has domain S and range T, and if for each element $t \in T$ there is some $s \in S$ such that $f(s) = t$ then we say that f is *onto*.

If the function g has domain S and range T, and if the only way that $f(s_1)$ can equal $f(s_2)$ is if $s_1 = s_2$ then we say that f is *one-to-one*.

EXAMPLE 1.3.3. Let $S = \{-3, -2, -1, 0, 1, 2, 3\}$ and let $T = \{0, 1, 4, 9\}$. Let the function f be given by $f(x) = x^2$. Then the set of all values of f, applied to elements of S, is $\{0, 1, 4, 9\}$. Therefore f is onto. However notice that $f(-2) = f(2) = 4$. Therefore the function f is not one-to-one. □

1.4 OPERATIONS ON FUNCTIONS

Let f and g be functions with domain S and range T. We define

- $[f + g](x) \equiv f(x) + g(x)$
- $[f - g](x) \equiv f(x) - g(x)$
- $[f \cdot g](x) \equiv f(x) \cdot g(x)$
- $\left[\frac{f}{g}\right](x) \equiv \frac{f(x)}{g(x)}$ provided that $g(x) \neq 0$

EXAMPLE 1.4.1. Let $f(x) = x^3 - x$ and $g(x) = x^4$. Then

$$[f + g](x) = x^3 - x + x^4 , \qquad [f - g](x) = x^3 - x - x^4 ,$$

$$[f \cdot g](x) = (x^3 - x) \cdot x^4 = x^7 - x^5 , \qquad \left[\frac{f}{g}\right](x) = \frac{x^3 - x}{x^4} . \qquad \square$$

If $f : S \to T$ and $g : T \to U$ then we may consider the function $g \circ f$ defined by

$$(g \circ f)(x) = g(f(x)) .$$

If $f : S \to T$ is both one-to-one and onto then we may define a function f^{-1} by the rule $f^{-1}(x) = y$ if and only if $f(y) = x$. We call f^{-1} the *inverse* of the function f. We have the essential properties

$$(f \circ f^{-1})(t) = t \ \ \forall t \in T$$

and

$$(f^{-1} \circ f)(s) = s \ \ \forall s \in S .$$

EXAMPLE 1.4.2. Let $f(x) = x^2 - 3x$ and $g(x) = x^3 + 1$. Then

$$(f \circ g)(x) = (x^3 + 1)^2 - 3 \cdot (x^3 + 1)$$

and

$$(g \circ f)(x) = (x^2 - 3x)^3 + 1 .$$

The function f is not one-to-one because $f(0) = f(3) = 0$. But $g : \mathbb{R} \to \mathbb{R}$ is both one-to-one and onto. We may solve the equation

$$(g \circ g^{-1})(x) = x$$

to find that

$$[g^{-1}(x)]^3 + 1 = x$$

or

$$g^{-1}(x) = \sqrt[3]{x - 1} . \qquad \square$$

1.5 NUMBER SYSTEMS

The most rudimentary number system is the *natural numbers*. These are the counting numbers $1, 2, 3, \ldots$, which are denoted by the symbol $\mathbb{N}$. Of the four standard arithmetic operations, the natural numbers are closed only under addition and multiplication.

The *integers* comprise both the positive and negative whole numbers and also 0. We denote this set by $\mathbb{Z}$. Of the four standard arithmetic operations, the integers are closed under addition, subtraction, and multiplication.

The *rational numbers* consist of all quotients of integers. Thus m/n is a rational number provided that $m, n \in \mathbb{Z}$ and $n \neq 0$. We denote the set of all rational numbers by $\mathbb{Q}$. The set of rational numbers is closed under all four of the standard arithmetic operations, except that division by 0 is not allowed.

EXAMPLE 1.5.1. The number 4 is a natural number. Of course it is also an integer. Writing it as $4 = 4/1$ we also see that this number is a rational number.

The number -6 is an integer. It is not a natural number. Writing it as $-6 = (-6)/1$, we also see that this number is a rational number.

The number $2/3$ is neither an integer nor a natural number. But it is a rational number. □

The number system of greatest interest to us is the real number system. It contains the rational numbers, and has several other interesting properties as well. We explore the real numbers in the next subsection.

1.5.1 THE REAL NUMBERS

The rational numbers are a *field*. This means that there are operations of addition ($+$) and multiplication ($\times$) that satisfy the usual laws of arithmetic. In addition the field $\mathbb{Q}$ satisfies certain properties of the ordering ($<$):

1. If $x, y, z \in \mathbb{Q}$ and $y < z$ then $x + y < x + z$.

2. If $x, y \in \mathbb{Q}$, $x > 0$, and $y > 0$ then $x \cdot y > 0$.

Thus $\mathbb{Q}$ is an *ordered field*.

The real numbers will be an ordered field containing the rationals and satisfying an additional completeness property. We formulate that property in terms of *least upper bound*.

Definition 1.5.2. Let $S \subseteq \mathbb{R}$. The set S is called *bounded above* if there is an element $b \in \mathbb{R}$ such that $x \leq b$ for all $x \in S$. We call the element b an *upper bound* for the set S.

Definition 1.5.3. Let $S \subseteq \mathbb{R}$. An element $b \in \mathbb{R}$ is called a *least upper bound* (or *supremum*) for S if b is an upper bound for S and there is no upper bound b' for S that is less than b. We write $b = \sup S = \text{lub}\, S$.

EXAMPLE 1.5.4. Let $S = \{x \in \mathbb{Q} : 3 < x < 5\}$. Then the number 9 is an upper bound for S, as is the number 7. The least upper bound for S is 5. We write $5 = \operatorname{lub} S = \sup S$. □

By its very definition, if a least upper bound exists then it is unique.

Before we go on, let us record a companion notion for lower bounds:

Definition 1.5.5. Let $S \subseteq \mathbb{R}$. The set S is called *bounded below* if there is an element $c \in \mathbb{R}$ such that $x \geq c$ for all $x \in S$. We call the element c a *lower bound* for the set S.

Definition 1.5.6. Let $S \subseteq \mathbb{R}$. An element $c \in \mathbb{R}$ is called a *greatest lower bound* (or *infimum*) for S if c is a lower bound for S and there is no lower bound c' for S that is greater than c. We write $c = \inf S = \operatorname{glb} S$.

By definition, if a greatest lower bound exists then it is unique.

EXAMPLE 1.5.7. Let $S = \{x \in \mathbb{R} : 0 < x < 1\}$ and $T = \{x \in \mathbb{R} : 0 \leq x < 1\}$. Then -1 is a lower bound both for S and for T and 0 is the greatest lower bound for both sets. We write $0 = \operatorname{glb} S$ and $0 = \operatorname{glb} T$. We may also write $0 = \inf S$ and $0 = \inf T$. Notice that $0 \notin S$ while $0 \in T$.

Also 5 is an upper bound both for S and for T, and 1 is the least upper bound for both sets. We write $1 = \operatorname{lub} S$ and $1 = \operatorname{lub} T$. We may also write $1 = \sup S$ and $1 = \sup T$. Observe that 1 is not in S and is not in T. □

Now we have:

Theorem 1.5.8. *There exists an ordered field* $\mathbb{R}$ *that* **(i)** *contains* $\mathbb{Q}$ *as a subfield and* **(ii)** *has the property that any non-empty subset of* $\mathbb{R}$ *that has an upper bound also has a least upper bound* (*that is also an element of* $\mathbb{R}$).

An equivalent, companion, statement is that if T is any set that is bounded below then T has a greatest lower bound (that is also an element of $\mathbb{R}$).

EXAMPLE 1.5.9. It is known (see [KRA2, page 114]) that there is no rational number whose square is 2—see Example 1.5.10 below. Let

$$S = \{x \in \mathbb{R} : x > 0 \text{ and } x^2 < 2\}.$$

Of course S is bounded above (by 2, for example), and so has least upper bound α. Of course α will be an element of $\mathbb{R}$, but $\alpha \notin \mathbb{Q}$. It can be shown that $\alpha^2 = 2$ (see [KRA1, Section 2.5, Theorem 12]). Thus the real number system contains numbers that are missing from the rational number system. These are called the *irrational numbers*. □

It can also be shown that the number π, which represents the ratio of the circumference of a circle to its diameter, is not a rational number. But π *does* exist as a real number.

EXAMPLE 1.5.10. Let us confirm that $\sqrt{2}$ is not a rational number. Suppose to the contrary that it is. So $\sqrt{2} = p/q$, with p and q integers. By division, we may suppose that p and q have no common divisors.

Thus

$$\left(\frac{p}{q}\right)^2 = 2\,.$$

Multiplying this out gives

$$2q^2 = p^2\,.$$

Since 2 divides the left side, we conclude that 2 divides the right side. So 2 divides p. Write $p = 2r$ for r an integer.

Thus we have

$$2q^2 = (2r)^2\,.$$

Simplifying gives

$$q^2 = 2r^2\,.$$

Since 2 divides the right side, we conclude that 2 divides the left side. So 2 divides q. We have shown that 2 divides p and also that 2 divides q. This contradicts the assumption that p and q have no common divisors.

We conclude that $\sqrt{2}$ cannot be rational. □

It is considerably more difficult to prove that π is irrational. We cannot treat the matter here, but see [NIV].

We shall learn below that the set of numbers $\mathbb{R} \setminus \mathbb{Q}$ (the irrational numbers) is much larger than $\mathbb{Q}$ itself. Thus "most" real numbers are not rational.

1.6 COUNTABLE AND UNCOUNTABLE SETS

Georg Cantor's theory of countable and uncountable sets, and more generally of many orders of infinity, is an integral part of any treatment of real analysis. What we give here is a summary. Complete treatments may be found in [KRA1, Section 1.8] and [KRA2, Section 5.8].

Two sets S and T are said to have the *same cardinality* if there is a one-to-one, onto function $\phi : S \to T$. We write card S = card T. In this context we refer to such a function ϕ as a *bijection*, or just an isomorphism.

The surprise is that some unlikely pairs of sets have the same cardinality. In particular, it is possible for $S \subseteq T$, $S \neq T$, and yet $\operatorname{card} S = \operatorname{card} T$.

EXAMPLE 1.6.1. Let $A = \{\heartsuit, \spadesuit, \clubsuit\}$, $B = \{\%, \&, \#\}$, and $C = \{1, 2\}$. Then

$$f : \begin{cases} \heartsuit \to \# \\ \spadesuit \to \% \\ \clubsuit \to \&. \end{cases}$$

is a bijection of A to B. This gives mathematical confirmation of the obvious fact that A and B have the same cardinality. We write $\operatorname{card} A = \operatorname{card} B$.

On the other hand, it is impossible to construct a bijection from A to C. So A and C do not have the same cardinality. □

EXAMPLE 1.6.2. Let $S = \{\ldots, -6, -4, -2, 0, 2, 4, 6, \ldots\}$ (the even integers) and let $T = \mathbb{Z}$. Then obviously $S \subseteq T$ but $S \neq T$. Yet $\phi(n) = n/2$ is an isomorphism of S to T. So $\operatorname{card} S = \operatorname{card} T$. □

If two sets have the same cardinality, then we think of them as having the same size. For finite sets, this idea coincides with our intuition: two sets have the same cardinality if and only if they have the same (finite) number of elements. But for infinite sets this says something new.

If a set S has the same cardinality as $\mathbb{N}$, the natural numbers, then we say that S is *countable*.

EXAMPLE 1.6.3. Let $S = \mathbb{Z}$, the integers, and let $T = \mathbb{N}$, the natural numbers. Define

$$f(n) = (-1)^{n+1} \cdot \lfloor j/2 \rfloor .$$

Here $\lfloor \ \rfloor$ is the greatest integer function. Then f is a bijection of T to S. So S and T have the same cardinality. We say that the integers are a countable set. □

EXAMPLE 1.6.4. Let $S = \{\ldots, -6, -4, -2, 0, 2, 4, 6, \ldots\}$. The last two examples show that S is countable. A similar argument shows that $T = \{\ldots, -5, -3, -1, 1, 3, 5, \ldots\}$ is countable. We will see below that the set $\mathbb{R}$ of real numbers is *not* countable. We say that $\mathbb{R}$ is *uncountable*. □

Cantor's great insight was that the set $\mathbb{R}$ of real numbers is in fact not countable (see [KRA1, Section 1.8] or [KRA2, Subsection 5.8.3] for a proof). If S is infinite and has cardinality different from the cardinality of $\mathbb{N}$ then we say that S is *uncountable*.

We now list some of the key properties of countable sets:

1. If both S and T are countable then $S \cup T$, $S \cap T$, and $S \times T$ are at most countable.
2. If X is uncountable and $Y \supseteq X$ then Y is uncountable.
3. If X is countable and $Y \subseteq X$, then Y is at most countable.

The phrase "at most countable" means either countably infinite or finite or empty. It is common to refer to a set of this type as "denumerable" (although many sources do not make this distinction very clearly). We reserve the word "countable" for infinite sets that have the same cardinality as the set $\mathbb{N}$.

EXAMPLE 1.6.5. The set $\mathbb{Q}$ can be identified in a natural way with a subset of $\mathbb{Z} \times \mathbb{Z}$, using the map $m/n \mapsto (m, n)$ (when m/n is in lowest terms). It follows that $\mathbb{Q}$ is countable. □

EXAMPLE 1.6.6. The set of all lines in the plane that contain (at least) two points having integer coefficients is countable. For each line may be identified with the 4-tuple of integers coming from the coordinates of the two given points, and the set of such 4-tuples is countable. □

EXAMPLE 1.6.7. The set $\mathbb{Z} \times \mathbb{R}$ is uncountable, for it contains a copy of $\mathbb{R}$ by way of the map

$$\mathbb{R} \ni x \mapsto (0, x) \in \mathbb{Z} \times \mathbb{R} .$$

□

EXAMPLE 1.6.8. The set $\mathbb{C}$, the complex numbers, is uncountable. It contains a copy of $\mathbb{R}$ by way of the map

$$\mathbb{R} \ni x \mapsto x + i0 \in \mathbb{C} .$$

□

EXAMPLE 1.6.9. Let us now construct a concrete example of an uncountable set. Our example will be the set S of all sequences on the set $\{0, 1\}$, i.e., the set of all infinite sequences of 0s and 1s. To see that S is uncountable, assume the contrary. Then there is a first sequence

$$\mathcal{S}^1 = \{s_j^{(1)}\}_{j=1}^{\infty} ,$$

a second sequence

$$\mathcal{S}^2 = \{s_j^{(2)}\}_{j=1}^{\infty} ,$$

and so forth. This will be a complete enumeration of all the members of S. But now consider the sequence $\mathcal{T} = \{t_j\}_{j=1}^{\infty}$, which we construct as follows:

- If $s_1^{(1)} = 0$ then set $t_1 = 1$; if $s_1^{(1)} = 1$ then set $t_1 = 0$;
- If $s_2^{(2)} = 0$ then set $t_2 = 1$; if $s_2^{(2)} = 1$ then set $t_2 = 0$;
- If $s_3^{(3)} = 0$ then set $t_3 = 1$; if $s_3^{(3)} = 1$ then set $t_3 = 0$;

$$\dots$$

- If $s_j^{(j)} = 0$ then set $t_j = 1$; if $s_j^{(j)} = 1$ then set $t_j = 0$;

etc.

Now the sequence $\mathcal{T}$ differs from the first sequence $\mathcal{S}^1$ in the first element: $t_1 \neq s_1^{(1)}$.

The sequence $\mathcal{T}$ differs from the second sequence $\mathcal{S}^2$ in the second element: $t_2 \neq s_2^{(2)}$.

And so on: the sequence $\mathcal{T}$ differs from the jth sequence $\mathcal{S}^j$ in the jth element: $t_j \neq s_j^{(j)}$. So the sequence $\mathcal{T}$ is not in the set S. But $\mathcal{T}$ is *supposed* to be in the set S because it is a sequence of 0s and 1s and all of these have been hypothesized to be enumerated.

This contradicts our assumption, so S must be uncountable. □

EXAMPLE 1.6.10. Consider the set of all decimal representations of numbers—both terminating and non-terminating. Here a terminating decimal is one of the form

$$27.43926$$

while a non-terminating decimal is one of the form

$$3.14159265\dots.$$

In a non-terminating decimal, no repetition is implied; the decimal simply continues without cease.

The set of all those decimals containing only the digits 0 and 1 can be identified in a natural way with the set of sequences containing only 0 and 1 (just put commas between the digits). We just saw that the set of such sequences is uncountable.

Since the set of all decimal numbers is an even bigger set, it must be uncountable also.

As you may know, the set of all decimals identifies with the set of all real numbers. We find then that the set $\mathbb{R}$ of all real numbers is uncountable. (Contrast this with the situation for the rationals.) □

It is an important result of set theory (due to Cantor) that, given any set S, the set of all subsets of S (called the *power set* of S) has strictly greater cardinality than the set S itself. As a simple example, let $S = \{a, b, c\}$. Then the set of all subsets of S is

$$\Big\{ \emptyset, \{a\}, \{b\}, \{c\}, \{a,b\}, \{a,c\}, \{b,c\}, \{a,b,c\} \Big\} .$$

The set of all subsets has eight elements while the original set has just three. In general, if S has k elements then the power set of S will have 2^k elements.

Even more significant is the fact that if S is an infinite set then the set of all its subsets has greater cardinality than S itself. This is a famous theorem of Cantor, implying that there are infinite sets of arbitrarily large cardinality.

We conclude this discussion with a result that makes it easy to determine the cardinalities of many sets.

Theorem 1.6.11 ([Schroeder-Bernstein]). *Let A, B be sets. If there is a one-to-one function $f : A \to B$ and a one-to-one function $g : B \to A$, then A and B have the same cardinality.*

CHAPTER 2

SEQUENCES

2.1 INTRODUCTION TO SEQUENCES

2.1.1 THE DEFINITION AND CONVERGENCE

Informally, a sequence is an ordered list of numbers:

$$a_1, a_2, a_3, \ldots.$$

In more formal treatments, we say that a *sequence* on a set S is a function f from $\mathbb{N}$ to S, and we identify $f(j)$ with a_j. Although the a_j (or $\{a_j\}_{j=1}^{\infty}$) notation is the most common, it is often useful to think of a sequence as a function.

EXAMPLE 2.1.1. Let

$$f(j) = 1/j^2.$$

This function defines the sequence

$$\frac{1}{1^2}, \frac{1}{2^2}, \frac{1}{3^2}, \ldots.$$

We also write

$$a_j = \frac{1}{j^2}.$$ □

The primary property of a sequence is its convergence or its non-convergence. We say that a sequence $\{a_j\}$ *converges* to a numerical limit ℓ if, for every $\epsilon > 0$, there is a positive integer N such that $j > N$ implies that $|a_j - \ell| < \epsilon$. We write $\lim_{j \to \infty} a_j = \ell$. Otherwise we say that the sequence *diverges*.

EXAMPLE 2.1.2. Consider the sequence $1, 1/2, 1/3, \ldots$. This sequence converges to 0. To see this, let $\epsilon > 0$ and choose N so large that $1/N < \epsilon$. If $j > N$, it follows that $|1/j - 0| = 1/j < 1/N < \epsilon$. Thus the sequence converges to 0. □

EXAMPLE 2.1.3. Consider the sequence $-1, 1, -1, 1, \ldots$. This sequence does not converge. We commonly say that it *diverges*. To see this, let $\epsilon = 1/2$. Denote the elements of the sequence by $a_j = (-1)^j$. Suppose that there were a limit ℓ and an $N > 0$ such that $j > N$ implies that $|a_j - \ell| < \epsilon = 1/2$. It follows that, for $j > N$, we have

$$\begin{aligned} 2 = |a_j - a_{j+1}| &= |(a_j - \ell) + (\ell - a_{j+1})| \\ &\leq |a_j - \ell| + |\ell - a_{j+1}| \\ &< \epsilon + \epsilon \\ &= 1 . \end{aligned}$$

The statement $2 < 1$ is false. So the limit ℓ does not exist and the sequence diverges. □

2.1.2 THE CAUCHY CRITERION

We can discuss the convergence or divergence of a sequence without making direct reference to the (putative) limit value ℓ. This is the significance of the *Cauchy criterion* or *Cauchy condition*. Let $\{a_j\}$ be a sequence. We say that the sequence satisfies the Cauchy criterion if, for each $\epsilon > 0$, there is an $N > 0$ such that, whenever $j, k > N$, then $|a_j - a_k| < \epsilon$. The Cauchy condition says, in effect, that the elements of the sequence are getting ever closer together (without making any statement about what point they may be getting *close to*). Put in other words, a sequence satisfies the Cauchy condition if the terms get so close together that, no matter how small a difference you have in mind, you can find an index after which all terms will be closer together than that. We sometimes say that a sequence satisfying this condition "is Cauchy".

EXAMPLE 2.1.4. Let $a_j = 1/2^j$. This sequence is Cauchy. For let $\epsilon > 0$ and choose N so large that $1/2^N < \epsilon$. Then, for $k > j > N$,

$$|a_j - a_k| < |a_j| < |a_N| = \frac{1}{2^N} < \epsilon .$$

Thus the sequence is Cauchy. □

The significance of the Cauchy criterion is given by the following result (which in turn hinges on the completeness property of the real numbers).

Proposition 2.1.5. *Let $\{a_j\}$ be a Cauchy sequence of real numbers. Then $\{a_j\}$ converges to an element of $\mathbb{R}$. Conversely, a convergent sequence in $\mathbb{R}$ satisfies the Cauchy criterion.*

The proof of this result involves a careful investigation of the completeness of the real number system, which we shall not treat here.

In particular, it follows from the last proposition that *any Cauchy sequence will have a limit in* $\mathbb{R}$.

The intuitive content of the Cauchy condition is that the elements of a Cauchy sequence *get close together and stay close together.* In other words, no matter how small a difference you have in mind, you can find an index after which all successive pairs will be closer than that distance. With this thought in mind, we readily see that the sequence in Example 2.1.3 cannot be Cauchy, so it cannot converge.

2.1.3 MONOTONICITY

Definition 2.1.6. Let $\{a_j\}$ be a sequence of real numbers. The sequence is said to be *monotone increasing* if $a_1 \leq a_2 \leq \dots$. It is *monotone decreasing* if $a_1 \geq a_2 \geq \dots$.

The word "monotone" is used here primarily for reasons of tradition. In many contexts the word is redundant and we omit it. We say that a sequence is *strictly monotone increasing* (resp. *strictly monotone decreasing*) if $a_j < a_{j+1}$ for every j (resp. $a_j > a_{j+1}$ for every j).

EXAMPLE 2.1.7. Let $a_j = 1/\sqrt{j}$. Then the sequence $\{a_j\}$ is monotone decreasing.

Let $b_j = (j-1)/j$. Then the sequence $\{b_j\}$ is monotone increasing. □

Proposition 2.1.8. *If $\{a_j\}$ is a monotone increasing sequence that is bounded above, so that $a_j \leq M < \infty$ for all j, then $\{a_j\}$ is convergent. If $\{a_j\}$ is a monotone decreasing sequence that is bounded below, so that $a_j \geq N > -\infty$ for all j, then $\{a_j\}$ is convergent.*

Corollary 2.1.9. *Let S be a set of real numbers that is bounded above and below, with β its supremum and α its infimum. If $\epsilon > 0$ then there are $s, t \in S$ such that $|s - \beta| < \epsilon$ and $|t - \alpha| < \epsilon$.*

This fact can now be construed in the language of sequences:

Corollary 2.1.10. *Let S be a set of real numbers that is bounded above and below. Let β be its supremum and α its infimum. There is a sequence $\{a_j\} \subseteq S$ and a sequence $\{b_j\} \subseteq S$ such that $a_j \to \alpha$ and $b_j \to \beta$.*

EXAMPLE 2.1.11. Let $S = \{x \in \mathbb{R} : 0 < x < 1\}$. The infimum of S is 0, and the sequence $a_j = 1/j \in S$ converges to 0. Likewise, the supremum of S is 1, and the sequence $b_j = (j-1)/j \in S$ converges to 1. □

2.1.4 THE PINCHING PRINCIPLE

We next turn to one of the most useful results for calculating the limit of a sequence:

Proposition 2.1.12 (The Pinching Principle). *Let $\{a_j\}$, $\{b_j\}$, and $\{c_j\}$ be sequences of real numbers satisfying*

$$a_j \leq b_j \leq c_j$$

for every j. If

$$\lim_{j\to\infty} a_j = \lim_{j\to\infty} c_j = \alpha$$

for some real number α, then

$$\lim_{j\to\infty} b_j = \alpha.$$

EXAMPLE 2.1.13. Let $a_j = [\sin j]/j$. Observe that

$$-\frac{1}{j} \leq a_j \leq \frac{1}{j}.$$

The two sequences between which $\{a_j\}$ is pinched obviously tend to zero (reference Example 2.1.2). Hence $\{a_j\}$ converges to 0. □

2.1.5 SUBSEQUENCES

Let $\{a_j\}$ be a sequence. If

$$0 < j_1 < j_2 < \dots$$

are positive integers then the function

$$k \mapsto a_{j_k}$$

is called a *subsequence* of the given sequence. We usually write the subsequence as

$$\{a_{j_k}\}_{k=1}^{\infty} \qquad \text{or} \qquad \{a_{j_k}\}.$$

Sometimes a sequence will be divergent, but will have a convergent subsequence.

EXAMPLE 2.1.14. Consider the sequence $a_j = (-1)^j$, as in Example 2.1.3. The subsequences $1, 1, 1, 1, \ldots$ and $-1, -1, -1, -1, \ldots$ are both convergent. □

A basic result about subsequences is this.

Proposition 2.1.15. *If $\{a_j\}$ is a convergent sequence with limit ℓ, then every subsequence converges to ℓ. Conversely, if $\{b_j\}$ is a sequence such that every subsequence converges to some limit m, then the full sequence converges to the limit m.*

EXAMPLE 2.1.16. Let $a_1 = 1/2$. Let a_2 be chosen so that $|a_2| < 1$, $1 - |a_2| < (1 - |a_1|)/2$, and with randomly selected sign. Inductively, choose a_{j+1} such that $|a_{j+1}| < 1$, $1 - |a_{j+1}| < (1 - |a_j|)/2$, and with randomly selected sign.

Then it easy to see that there is either a monotone increasing subsequence or a monotone decreasing subsequence of the a_j. The full sequence $\{a_j\}$ will, in general, not converge. But the indicated monotone subsequence is bounded in absolute value by 1 so will converge by Proposition 2.1.8. □

2.1.6 THE BOLZANO-WEIERSTRASS THEOREM

The fundamental theorem about the existence of convergent subsequences is this:

Theorem 2.1.17 (Bolzano-Weierstrass). *Let $\{a_j\}$ be a bounded sequence. Then there is a convergent subsequence $\{a_{j_k}\}$.*

EXAMPLE 2.1.18. We know that the set $\mathbb{Q}$ of rational numbers in the unit interval $[0, 1]$ is countable. Let them be enumerated as $\{a_1, a_2, \ldots\}$. This sequence will be a quite chaotic subset of the unit interval. Nevertheless, the Bolzano-Weierstrass theorem guarantees that it has a convergent subsequence.

Likewise, the sequence $a_j = \sin j$ is bounded. If you write out the first ten or twenty terms (use your calculator), you will see that this, too, is a rather unpredictable sequence. But the theorem guarantees the existence of a convergent subsequence. □

2.2 LIMSUP AND LIMINF

Let $\{a_j\}$ be any sequence of real numbers. The *limit supremum* of this sequence is the greatest limit of all subsequences of the given sequence. More rigorously, for each j let

$$A_j = \sup\{a_j, a_{j+1}, a_{j+2}, \ldots\}.$$

Then $\{A_j\}$ is a monotone decreasing sequence (since as j becomes large we are taking the supremum of a smaller set of numbers), so it has a limit. We define the limit supremum of $\{a_j\}$ to be

$$\limsup a_j = \lim_{j\to\infty} A_j .$$

The limit supremum may be $\pm\infty$.

Likewise, the *limit infimum* of the given sequence is the least limit of all subsequences of the given sequence. In detail, let

$$B_j = \inf\{a_j, a_{j+1}, a_{j+2}, \ldots\}.$$

Then $\{B_j\}$ is a monotone increasing sequence (since as j becomes large we are taking the infimum of a smaller set of numbers), so it has a limit. We define the limit infimum of $\{a_j\}$ to be

$$\liminf a_j = \lim_{j\to\infty} B_j ,$$

which also may be $\pm\infty$.

EXAMPLE 2.2.1. The sequence $a_j = (-1)^j$ has limit supremum 1 and limit infimum -1.

It is less obvious, but true, that the limit supremum of the sequence $\{\sin j\}$ is 1 and the limit infimum of this sequence is -1. □

The following result is now intuitively obvious, but worth noting explicitly.

Proposition 2.2.2. *Let $\{a_j\}$ be a sequence and set* $\limsup a_j = \beta$ *and* $\liminf a_j = \alpha$. *Assume that α, β are finite real numbers. Let $\epsilon > 0$. Then there are arbitrarily large j such that $a_j > \alpha - \epsilon$ and arbitrarily large k such that $a_k < \beta + \epsilon$. Compare Corollaries 2.1.9, 2.1.10.*

EXAMPLE 2.2.3. Let $a_j = \sin j$. A calculator calculation indicates that the limit supremum of $\{a_j\}$ is 1 and the limit infimum is -1. In the course of calculating with your handheld, you will have produced elements of the sequence that are arbitrarily near to -1, and you will also have produced elements that are arbitrarily near to $+1$. Thus your calculations illustrate the proposition. A rigorous proof of this result requires *Weyl's lemma* (see [STR]). □

We conclude this brief consideration of lim sup and lim inf with a result that ties all the ideas together.

Proposition 2.2.4. *Let $\{a_j\}$ be a sequence of real numbers. We define $\beta = \limsup_{j\to\infty} a_j$ and $\alpha = \liminf_{j\to\infty} a_j$. If $\{a_{j_k}\}$ is any subsequence of the given sequence then*

$$\alpha \le \liminf_{k\to\infty} a_{j_k} \le \limsup_{k\to\infty} a_{j_k} \le \beta\,.$$

Moreover, there is a subsequence $\{a_{j_l}\}$ such that

$$\lim_{l\to\infty} a_{j_l} = \alpha$$

and another subsequence $\{a_{j_m}\}$ such that

$$\lim_{m\to\infty} a_{j_m} = \beta\,.$$

Again, compare Corollary 2.1.10.

EXAMPLE 2.2.5. Let

$$a_j = j\cdot\pi - \left[\text{the greatest integer not exceeding } j\cdot\pi\right].$$

Every element of $\{a_j\}$ lies between 0 and 1, and none is equal to 0 or 1. You can use your calculator to convince yourself that there are elements of the sequence that are arbitrarily near to 0 and other elements that are arbitrarily near to 1. We may say that the limit supremum of the sequence is 1 and the limit infimum of the sequence is 0. Thus you will see empirically that there is a subsequence converging to 0, and another subsequence converging to 1. Again, Weyl's lemma [STR] can be used to give a rigorous treatment of these ideas. □

2.3 SOME SPECIAL SEQUENCES

It is useful to have a collection of special sequences for comparison and study.

EXAMPLE 2.3.1. Fix a real number λ. The sequence $\{\lambda^j\}$ is called a *power sequence*. If $-1 < \lambda < 1$ then the sequence converges to 0. If $\lambda = 1$ then the sequence is a constant sequence and converges to 1. If $\lambda > 1$ then the sequence diverges to $+\infty$. Finally, if $\lambda \leq -1$ then the sequence diverges.

□

For $\alpha > 0$, we define

$$\alpha^{m/n} = (\alpha^m)^{1/n},$$

where n is a positive integer and $m \in \mathbb{Z}$. Here the nth root (i.e., $(1/n)^{\text{th}}$ power) of a positive number is defined just like the square root was in Example 1.5.9. Thus we may talk about rational powers of a positive number. If $\beta \in \mathbb{R}$, then we may define

$$\alpha^{\beta} = \sup\{\alpha^{q} : q \in \mathbb{Q}, q < \beta\}.$$

Thus we can define any real power of a positive real number.

Lemma 2.3.2. *If $\alpha > 1$ is a real number and $\beta > 0$ then $\alpha^{\beta} > 1$.*

EXAMPLE 2.3.3. Fix a real number α and consider the sequence $\{j^{\alpha}\}$. If $\alpha > 0$ then it is easy to see that $j^{\alpha} \to +\infty$: to verify this assertion fix $M > 0$ and take the number N to be the first integer after $M^{1/\alpha}$.

If $\alpha = 0$ then j^{α} is a constant sequence, identically equal to 1.

If $\alpha < 0$ then $j^{\alpha} = 1/j^{|\alpha|}$. The denominator of this last expression tends to $+\infty$ hence the sequence $\{j^{\alpha}\}$ tends to 0. □

EXAMPLE 2.3.4. The sequence $\{j^{1/j}\}$ converges to 1. In fact, consider the expressions $\alpha_j = j^{1/j} - 1 > 0$. We have (by the Binomial Theorem) that

$$j = (\alpha_j + 1)^j \geq \frac{j(j-1)}{2}(\alpha_j)^2 .$$

Thus

$$0 < \alpha_j \leq \sqrt{2/(j-1)}$$

as long as $j \geq 2$. It follows from Proposition 2.1.12 that $\alpha_j \to 0$ or $j^{1/j} \to 1$. □

EXAMPLE 2.3.5. Let α be a positive real number. Then the sequence $\{\alpha^{1/j}\}$ converges to 1. To see this, first note that the case $\alpha = 1$ is trivial, and the case $\alpha > 1$ implies (by taking reciprocals) the case $\alpha < 1$. So we concentrate on $\alpha > 1$. Then we have

$$1 < \alpha^{1/j} < j^{1/j}$$

when $j > \alpha$. Since $j^{1/j}$ tends to 1, Proposition 2.1.12 applies and the argument is complete. □

EXAMPLE 2.3.6. Let $\lambda > 1$ and let α be real. Then the sequence

$$\left\{\frac{j^\alpha}{\lambda^j}\right\}_{j=1}^\infty$$

converges to 0.

To see this, fix an integer $k > \alpha$ and consider $j > 2k$. (Notice that k is fixed once and for all but j will be allowed to tend to $+\infty$ at the appropriate moment.) Writing $\lambda = 1 + \mu$, $\mu > 0$, we have that

$$\lambda^j = (\mu+1)^j > \frac{j(j-1)(j-2)\cdots(j-k+1)}{k(k-1)(k-2)\cdots 2\cdot 1}\mu^k \cdot 1^{j-k}.$$

This comes from picking out the kth term of the binomial expansion for $(\mu+1)^j$. Since $j > 2k$, each of the expressions $j, (j-1), \ldots, (j-k+1)$ in the numerator on the right exceeds $j/2$. Thus

$$\lambda^j > \frac{j^k}{2^k \cdot k!}\cdot \mu^k \quad \text{and} \quad 0 < \frac{j^\alpha}{\lambda^j} < j^\alpha \cdot \frac{2^k \cdot k!}{j^k \cdot \mu^k} = \frac{j^{\alpha-k}\cdot 2^k \cdot k!}{\mu^k}.$$

Since $\alpha - k < 0$, the right side tends to 0 as $j \to \infty$. □

EXAMPLE 2.3.7. The sequence

$$\left\{\left(1+\frac{1}{j}\right)^j\right\}_{j=1}^\infty$$

converges. In fact it is monotone increasing and bounded above. Use the Binomial Expansion to verify this assertion. The limit of the sequence is the number that we shall later call e (in honor of Leonhard Euler, 1707–1783, who first studied it in detail). We shall study this sequence further in Section 3.4. □

EXAMPLE 2.3.8. The sequence

$$\left\{\left(1-\frac{1}{j}\right)^j\right\}_{j=1}^\infty$$

converges to $1/e$, where the definition of e is given in the last example. More generally, the sequence

$$\left\{1+\frac{x}{j}\right\}^j$$

converges to e^x (here e^x is defined as in the discussion following Example 2.3.1 above). □

CHAPTER 3

Series

3.1 Introduction to Series

3.1.1 The Definition and Convergence

A *series* is, informally speaking, an infinite sum. We write a series as

$$\sum_{j=1}^{\infty} c_j .$$

We think of the series as meaning

$$\sum_{j=1}^{\infty} c_j = c_1 + c_2 + c_3 + \cdots .$$

The basic question about a series is "Does the series converge?" That is to say, does the infinite sum have any meaning? Does it represent some finite real number?

EXAMPLE 3.1.1. Consider the series

$$\sum_{j=1}^{\infty} \frac{1}{3^j} .$$

Although we do not yet know the rigorous ideas connected with series, we may think about this series heuristically. We may consider the "sum" of this series by adding together finitely many of its terms:

$$S_N = \sum_{j=1}^{N} \frac{1}{3^j} .$$

It is easy to calculate that $S_N = \frac{1}{2}(1 - 3^{-N})$. Thus the limit as N tends to ∞ of S_N is $1/2$. We intuit therefore that the sum of this series is $1/2$. The theory presented below will confirm this calculation. □

3.1.2 PARTIAL SUMS

With a view to answering our fundamental question, we define the *partial sum* of the series $\sum_{j=1}^{\infty} c_j$ to be

$$S_N \equiv c_1 + c_2 + \cdots + c_N .$$

We say that the series *converges* if the sequence of partial sums $\{S_N\}$ converges to a finite limit.

EXAMPLE 3.1.2. Let $c_j = 2^{-j}$. Then the Nth partial sum of $\sum_{j=1}^{\infty} c_j$ is

$$\begin{aligned} S_N &= 2^{-1} + 2^{-2} + \cdots 2^{-N} \\ &= \left(1 - 2^{-1}\right) + \left(2^{-1} - 2^{-2}\right) + \cdots + \left(2^{-N+1} - 2^{-N}\right) \\ &= 1 - 2^{-N} . \end{aligned}$$

We see that

$$\begin{aligned} \lim_{N\to\infty} S_N &= \lim_{N\to\infty} 1 - 2^{-N} \\ &= 1 . \end{aligned}$$

Thus the limit of the partial sums exists and the series converges. □

The series in the last example—in which each successive term is the product of the preceding term with a fixed constant—is commonly known as a *geometric series.*

EXAMPLE 3.1.3. Let $c_j = (-1)^j$. Then the sequence of partial sums is

$$-1, 0, -1, 0, \ldots .$$

It is plain that this sequence has no limit. So the series does not converge. □

EXAMPLE 3.1.4. Let $c_j = 1/j$. Then

$$\begin{aligned} S_2 &= 1 + \frac{1}{2} \\ S_4 &= 1 + \frac{1}{2} + \frac{1}{3} + \frac{1}{4} = 1 + \left[\frac{1}{2}\right] + \left[\frac{1}{3} + \frac{1}{4}\right] \\ &> 1 + \left[\frac{1}{2}\right] + \left[\frac{1}{4} + \frac{1}{4}\right] = 1 + \frac{1}{2} + \frac{1}{2} \end{aligned}$$

$$
\begin{aligned}
S_8 &= 1 + \frac{1}{2} + \frac{1}{3} + \frac{1}{4} + \frac{1}{5} + \frac{1}{6} + \frac{1}{7} + \frac{1}{8} \\
&= 1 + \left[\frac{1}{2}\right] + \left[\frac{1}{3} + \frac{1}{4}\right] + \left[\frac{1}{5} + \frac{1}{6} + \frac{1}{7} + \frac{1}{8}\right] \\
&> 1 + \frac{1}{2} + \left[\frac{1}{4} + \frac{1}{4}\right] + \left[\frac{1}{8} + \frac{1}{8} + \frac{1}{8} + \frac{1}{8}\right] \\
&= 1 + \frac{1}{2} + \frac{1}{2} + \frac{1}{2}
\end{aligned}
$$

etc.

We see that $S_j > (j+2)/2$ hence the sequence $S_1, S_2, \ldots$ of partial sums is strictly increasing, and it has a subsequence that tends to $+\infty$. Thus the sequence of partial sums does not tend to a finite limit, and the series diverges. □

The series in the last example is commonly known as the *harmonic series*.

3.2 ELEMENTARY CONVERGENCE TESTS

3.2.1 THE COMPARISON TEST

Proposition 3.2.1. *Suppose that $\sum_{j=1}^{\infty} c_j$ is a convergent series of non-negative terms. If $\{b_j\}$ is a sequence of real numbers, and if $|b_j| \leq c_j$ for every j, then the series $\sum_{j=1}^{\infty} b_j$ converges.*

Corollary 3.2.2. *If $\sum_{j=1}^{\infty} c_j$ is as in the proposition and if $0 \leq b_j \leq c_j$ for every j then the series $\sum_{j=1}^{\infty} b_j$ converges.*

EXAMPLE 3.2.3. The series $\sum_{j=1}^{\infty} 2^{-j} \sin j$ is seen to converge by comparing it with the series $\sum_{j=1}^{\infty} 2^{-j}$. □

EXAMPLE 3.2.4. The series $\sum_{j=1}^{\infty} \ln j/3^j$ is seen to converge by comparing it with the series $\sum_{j=1}^{\infty} 1/2^j$. □

3.2.2 THE CAUCHY CONDENSATION TEST

Theorem 3.2.5 (Cauchy Condensation Test). *Assume that $c_1 \geq c_2 \geq \cdots \geq c_j \geq \ldots 0$. The series*

$$\sum_{j=1}^{\infty} c_j$$

converges if and only if the series

$$\sum_{k=1}^{\infty} 2^k \cdot c_{2^k}$$

converges.

EXAMPLE 3.2.6. We apply the Cauchy condensation test to the harmonic series

$$\sum_{j=1}^{\infty} \frac{1}{j}.$$

It leads us to examine the series

$$\sum_{k=1}^{\infty} 2^k \cdot \frac{1}{2^k} = \sum_{k=1}^{\infty} 1.$$

Since the latter series diverges, the harmonic series diverges as well. □

EXAMPLE 3.2.7. The series

$$\sum_{j=1}^{\infty} \frac{1}{j^r}$$

converges if r is a real number that exceeds 1 and diverges if $r \leq 1$. We leave the details as an exercise for the reader. Use the Cauchy test.

3.2.3 GEOMETRIC SERIES

Proposition 3.2.8. *Let α be a fixed real number. The series*

$$\sum_{j=0}^{\infty} \alpha^j$$

is called a geometric series. It is useful to write

$$S_N = 1 + \alpha + \alpha^2 + \cdots + \alpha^{N-1} + \alpha^N$$

hence

$$\alpha \cdot S_N = \alpha + \alpha^2 + \alpha^3 + \cdots + \alpha^N + \alpha^{N+1} = S_N + \alpha^{N+1} - 1\,.$$

Thus

$$S_N = \frac{1 - \alpha^{N+1}}{1 - \alpha}\,.$$

It follows that the series converges if and only if $|\alpha| < 1$. In this circumstance, the sum of the series (that is, the limit of the partial sums) is $1/(1 - \alpha)$.

We already examined particular geometric series in Example 3.1.1 and 3.1.2.

EXAMPLE 3.2.9. The series $\sum_j (3.1)^{-j}$ is a geometric series with partial sums

$$S_N = \frac{1 - 3.1^{-(N+1)}}{1 - 3.1^{-1}} .$$

The series converges to

$$S = \frac{31}{21} .$$

□

3.2.4 THE ROOT TEST

Theorem 3.2.10. *If*

$$\limsup_{j \to \infty} |c_j|^{1/j} < 1 ,$$

then the series $\sum_{j=1}^{\infty} c_j$ *converges.*

3.2.5 THE RATIO TEST

Theorem 3.2.11. *If*

$$\limsup_{j \to \infty} \left| \frac{c_{j+1}}{c_j} \right| < 1 ,$$

then the series $\sum_{j=1}^{\infty} c_j$ *converges.*

Remark 3.2.12. If a series passes the Ratio Test then it passes the Root Test, but not conversely. That is, the Root Test is a better test than the Ratio Test because it will give information whenever the Ratio Test does and also in some circumstances when the Ratio Test does not.

Why do we therefore learn the Ratio Test? The answer is that there are circumstances when the Ratio Test is much easier to apply than the Root Test.

EXAMPLE 3.2.13. The series

$$\sum_{j=1}^{\infty} \frac{2^j}{j!}$$

is easily studied using the Ratio Test (recall that $j! \equiv j \cdot (j-1) \cdot \ldots 2 \cdot 1$). Indeed $c_j = 2^j/j!$ and

$$\left| \frac{c_{j+1}}{c_j} \right| = \frac{2^{j+1}/(j+1)!}{2^j/j!} .$$

We can perform the division to see that

$$\left|\frac{c_{j+1}}{c_j}\right| = \frac{2}{j+1}.$$

The lim sup of the last expression is 0. By the Ratio Test, the series converges.

In this example, while the Root Test applies in principle, it would be difficult to use in practice. □

EXAMPLE 3.2.14. We apply the Root Test to the series

$$\sum_{j=1}^{\infty} \frac{j^2}{2^j}$$

Observe that

$$c_j = \frac{j^2}{2^j}$$

hence that

$$|c_j|^{1/j} = \frac{(j^{1/j})^2}{2}.$$

As $j \to \infty$, we see that

$$\limsup_{j\to\infty} |c_j|^{1/j} = \frac{1}{2} < 1\,.$$

By the Root Test, the series converges. □

3.2.6 Root and Ratio Tests for Divergence

It is natural to ask whether the Ratio and Root Tests can detect divergence. Neither test is necessary and sufficient: there are series that elude the analysis of both tests. We still have these useful results:

Theorem 3.2.15 (Ratio Test for Divergence). *If there is an $N > 0$ such that*

$$\left|\frac{c_{j+1}}{c_j}\right| \geq 1\ , \quad \forall j \geq N\,,$$

then the series $\sum_{j=1}^{\infty} c_j$ diverges.

Theorem 3.2.16 (Root Test for Divergence). *If*

$$\limsup_{j\to\infty} |c_j|^{1/j} > 1\,,$$

then the series $\sum_{j=1}^{\infty} c_j$ diverges.

In the Ratio or Root Tests, if the lim sup is 1, then no conclusion is possible.

EXAMPLE 3.2.17. Consider the series

$$\sum_{j=1}^{\infty} \frac{j^{j/2}}{3^j} .$$

Setting $c_j = j^{j/2}/3^j$, we calculate that

$$\lim_{j\to\infty} |c_j|^{1/j} = \lim_{j\to\infty} j^{1/2}/3 = +\infty .$$

By Theorem 3.2.16, the series diverges.

Now consider the series

$$\sum_{j=1}^{\infty} \frac{1}{j^2} .$$

If we set $c_j = 1/j^2$, then we see that

$$\lim_{j\to\infty} |c_j|^{1/j} = \lim_{j\to\infty} \frac{1}{[j^{1/j}]^2} = 1 .$$

The Root Test therefore gives us no information. However, we can use the Cauchy Condensation Test to see that the series converges. See also Example 3.2.7. □

EXAMPLE 3.2.18. For the series

$$\sum_{j=1}^{\infty} \frac{j!}{4^j} ,$$

setting $c_j = j!/4^j$, we calculate that

$$\lim_{j\to\infty} |c_{j+1}/c_j| = \lim_{j\to\infty} [j+1]/4 = +\infty .$$

By Theorem 3.2.15, the series diverges.

For the series

$$\sum_{j=1}^{\infty} \frac{1}{j} ,$$

set $c_j = 1/j$. Then we see that

$$\lim_{j\to\infty} \left| \frac{c_{j+1}}{c_j} \right| = \lim_{j\to\infty} \frac{j}{j+1} = 1 .$$

The Ratio Test therefore gives us no information. However, we can use the Cauchy Condensation Test, as we saw in Example 3.2.6, to see that the series diverges. See also Example 3.1.4. □

3.3 Advanced Convergence Tests

3.3.1 Summation by Parts

In this section we consider convergence tests for series that depend on cancellation among their terms.

Proposition 3.3.1. *Let* $\{a_j\}_{j=0}^{\infty}$ *and* $\{b_j\}_{j=0}^{\infty}$ *be two sequences of real or complex numbers. For* $N = 0, 1, 2, \ldots$ *set*

$$A_N = \sum_{j=0}^{N} a_j$$

(we adopt the convention that $A_{-1} = 0$.*) Then for any* $0 \leq m \leq n < \infty$ *it holds that*

$$\begin{aligned}\sum_{j=m}^{n} a_j \cdot b_j &= [A_n \cdot b_n - A_{m-1} \cdot b_m] \\ &\quad + \sum_{j=m}^{n-1} A_j \cdot (b_j - b_{j+1}).\end{aligned}$$

3.3.2 Abel's Test

Summation by parts may be used to derive the following test of Niels Henrik Abel (1802–1829).

Theorem 3.3.2 (Abel). *Consider the series*

$$\sum_{j=0}^{\infty} a_j \cdot b_j.$$

If

1. *the partial sums* $A_N = \sum_{j=0}^{N} a_j$ *form a bounded sequence,*

2. $b_0 \geq b_1 \geq b_2 \geq \ldots,$

3. $\lim_{j\to\infty} b_j = 0,$

then the original series

$$\sum_{j=0}^{\infty} a_j \cdot b_j$$

converges.

EXAMPLE 3.3.3. As a first application of Abel's convergence test, we examine alternating series. Consider a series of the form

$$\sum_{j=1}^{\infty}(-1)^j \cdot b_j, \tag{3.3.3.1}$$

with $b_1 \geq b_2 \geq b_3 \geq \cdots \geq 0$ and $b_j \to 0$ as $j \to \infty$. We set $a_j = (-1)^j$ and apply Abel's test. We see immediately that all partial sums A_N are either -1 or 0. So the sequence of partial sums is bounded. And the b_j's are monotone decreasing and tending to zero. By Abel's convergence test, the alternating series (3.3.3.1) converges. □

Proposition 3.3.4. *Let $b_1 \geq b_2 \geq \cdots \geq 0$ and assume that $b_j \to 0$. Consider the alternating series $\sum_{j=1}^{\infty}(-1)^j b_j$ as in the last example. It converges to a sum S. Then the partial sums S_N satisfy $|S - S_N| \leq b_{N+1}$.*

EXAMPLE 3.3.5. The series

$$\sum_{j=1}^{\infty}(-1)^j \frac{1}{j}$$

converges by Example 3.3.3. Then the partial sum $S_{100} = -.688172$ is within 0.01 (in fact within 1/101) of the full sum S and the partial sum $S_{10000} = -.6930501$ is within 0.0001 (in fact within 1/10001) of S. □

EXAMPLE 3.3.6. Next we examine a series that is important in the study of Fourier analysis. Consider the series

$$\sum_{j=1}^{k} \frac{\sin j}{j}. \tag{3.3.6.1}$$

We already know that the series $\sum \frac{1}{j}$ diverges. However, the expression $\sin j$ changes sign in a rather sporadic fashion. We might hope that the series (3.3.6.1) converges because of cancellation of the summands. We take $a_j = \sin j$ and $b_j = 1/j$. Abel's test will apply if we can verify that the partial sums A_N of the a_j's are bounded. To see this we use a trick:

We know that

$$\cos(j + 1/2) = \cos j \cdot \cos 1/2 - \sin j \cdot \sin 1/2$$

and

$$\cos(j - 1/2) = \cos j \cdot \cos 1/2 + \sin j \cdot \sin 1/2.$$

Subtracting these equations and solving for $\sin j$ yields

$$\sin j = \frac{\cos(j - 1/2) - \cos(j + 1/2)}{2 \cdot \sin 1/2}.$$

We conclude that

$$A_N = \sum_{j=1}^{N} a_j = \sum_{j=1}^{N} \frac{\cos(j - 1/2) - \cos(j + 1/2)}{2 \cdot \sin 1/2}.$$

This sum collapses and we see that

$$A_N = \frac{-\cos(N + 1/2) + \cos 1/2}{2 \cdot \sin 1/2}.$$

Thus

$$|A_N| \leq \frac{2}{2 \cdot \sin 1/2} = \frac{1}{\sin 1/2},$$

independent of N.

Thus the hypotheses of Abel's test are verified and the series

$$\sum_{j=1}^{k} \frac{\sin j}{j}$$

is seen to converge. □

Remark 3.3.7. It is interesting that both the series

$$\sum_{j=1}^{k} \frac{|\sin j|}{j} \quad \text{and} \quad \sum_{j=1}^{k} \frac{\sin^2 j}{j}$$

diverge. The details of these assertions are left to the reader.

3.3.3 ABSOLUTE AND CONDITIONAL CONVERGENCE

We turn next to the topic of absolute and conditional convergence. A series of real or complex constants

$$\sum_{j=1}^{\infty} a_j$$

is said to be *absolutely convergent* if

$$\sum_{j=1}^{\infty} |a_j|$$

converges. We have:

Proposition 3.3.8. *If the series* $\sum_{j=1}^{k} a_j$ *is absolutely convergent then it is convergent.*

Definition 3.3.9. A series $\sum_{j=1}^{k} a_j$ is said to be *conditionally convergent* if $\sum_{j=1}^{k} a_j$ converges, but does not converge absolutely.

Absolutely convergent series are convergent but the next example shows that the converse is not true.

EXAMPLE 3.3.10. The series

$$\sum_{j=1}^{k} \frac{(-1)^j}{j} \tag{3.3.10.1}$$

converges by the alternating series test. However, it is not absolutely convergent because the harmonic series

$$\sum_{j=1}^{\infty} \frac{1}{j}$$

diverges. Thus the series (3.3.10.1) is conditionally convergent. □

Remark 3.3.11. We know from Example 3.3.6 that the series

$$\sum_{j=1}^{\infty} \frac{\sin j}{j}$$

converges. Its terms vary in sign in a fairly erratic fashion. The cancellation is very subtle—this series is not an alternating series. As we have already noted—Remark 3.3.7—the series $\sum_j |\sin j|/j$ does *not* converge; so this series is conditionally convergent.

3.3.4 REARRANGEMENTS OF SERIES

There is a remarkable robustness result for absolutely convergent series that fails dramatically for conditionally convergent series. This result is enunciated in the next theorem. We first need a definition.

Definition 3.3.12. Let $\sum_{j=1}^{\infty} c_j$ be a given series. Let $\{p_j\}_{j=1}^{\infty}$ be a sequence in which every positive integer occurs once and only once (but not necessarily in the usual order). Then the series

$$\sum_{j=1}^{\infty} c_{p_j}$$

is said to be a *rearrangement* of the given series.

Theorem 3.3.13 (Weierstrass). *If the series $\sum_{j=1}^{\infty} a_j$ of real numbers is absolutely convergent then it is convergent; let the sum be ℓ. Then every rearrangement of the series converges also to ℓ.*

If the series $\sum_{j=1}^{\infty} b_j$ is conditionally convergent and if β is any real number or $\pm\infty$ then there is a rearrangement of the series such that its sequence of partial sums converges to β.

EXAMPLE 3.3.14. The series

$$\sum_{j=1}^{\infty} \frac{(-1)^j}{j}$$

is conditionally convergent (because it is an alternating series). By Weierstrass's theorem, there is a rearrangement of the series that converges to 5. How can we find it?

The series consisting of all the positive terms of the series will diverge (exercise). Likewise, the series consisting of all the negative terms of the series will diverge. Thus we construct the desired rearrangement by using the following steps:

(1) First select just enough positive terms to obtain a partial sum that is greater than 5.

(2) Then add on enough negative terms so that the partial sum falls below 5.

(3) Now add on enough positive terms so that the partial sum once again exceeds 5.

(4) Again add on enough negative terms so that the partial sum falls below 5.

Now continue in this fashion.

Because the series of positive terms diverges, Steps **(1)** and **(3)** (and subsequent odd-numbered) steps are possible. Because the series of negative terms diverges, Steps **(2)** and **(4)** (and subsequent even-numbered steps) are possible. Because the series converges conditionally, the terms of the series tend to zero. So the partial sums we are constructing are getting ever closer together. In sum, the construction yields a rearrangement that converges to 5. □

3.4 SOME PARTICULAR SERIES

3.4.1 THE SERIES FOR e

We begin with a series that defines a special constant of mathematical analysis.

Definition 3.4.1. The series

$$\sum_{j=0}^{\infty} \frac{1}{j!},$$

where $j! \equiv j \cdot (j-1) \cdot (j-2) \cdots 1$ for $j \geq 1$ and $0! \equiv 1$, is convergent (by the Ratio Test, for instance). Its sum is denoted by the symbol e in honor of the Swiss mathematician Leonhard Euler, who first studied it.

Like the number π, to be considered later, the number e is one that arises repeatedly in a variety of contexts in mathematics. It has many special properties. The first of these that we shall consider is that the definition that we have given for e is equivalent to another involving a sequence, considered earlier in Examples 2.3.7 and 2.3.8.

3.4.2 OTHER REPRESENTATIONS FOR e

Proposition 3.4.2. *The limit*

$$\lim_{n\to\infty} \left(1 + \frac{1}{n}\right)^n$$

exists and equals e.

We have already noted this fact in Example 2.3.7.

The next result tells us how rapidly the partial sums $A_N \equiv \sum_{j=0}^{N} \frac{1}{j!}$ of the series converge to e. This fact is of theoretical interest, and can be applied to demonstrate the irrationality of e.

Proposition 3.4.3. *If*

$$A_N = \sum_{j=0}^{N} \frac{1}{j!},$$

then

$$0 < e - A_N < \frac{1}{N \cdot N!}.$$

With some sharp theoretical work, the last estimate can be used to establish the following:

Theorem 3.4.4. *Euler's number e is irrational.*

For a reference, see [NIV], [RUD], or [KRA1].

3.4.3 Sums of Powers

It is part of mathematical legend that Carl Friedrich Gauss (1777–1855) was given the task, as a child, to sum the integers from 1 to 100. The story is that he derived a remarkable formula and gave the correct answer in a few moments. It is said that he reasoned as follows: Let $S = 1+2+\cdots+99+100$. Then

$$\begin{array}{rcccccccccccc} S & = & 1 & + & 2 & + & 3 & + \cdots + & 98 & + & 99 & + & 100 \\ S & = & 100 & + & 99 & + & 98 & + \cdots + & 3 & + & 2 & + & 1\,. \end{array}$$

Adding vertically, we find that

$$2S = \underbrace{101 + 101 + 101 + \cdots + 101 + 101 + 101}_{100 \text{ times}}\,.$$

Thus

$$2S = 100 \cdot 101 = 10100$$

and so

$$S = 5050\,.$$

The same reasoning may be used to show that

$$S_{1,N} \equiv \sum_{j=1}^{N} j = \frac{N(N+1)}{2}\,.$$

It is frequently of interest to sum higher powers of j. Say that we wish to calculate

$$S_{k,N} \equiv \sum_{j=1}^{N} j^k$$

for some positive integer k exceeding 1. We may proceed as follows: write

$$\begin{aligned} (j+1)^{k+1} - j^{k+1} &= \left[j^{k+1} + (k+1)\cdot j^k + \frac{(k+1)\cdot k}{2}\cdot j^{k-1} \right. \\ &\qquad \left. + \cdots + \frac{(k+1)\cdot k}{2}\cdot j^2 + (k+1)\cdot j + 1 \right] \\ &\qquad - j^{k+1} \\ &= (k+1)\cdot j^k + \frac{(k+1)\cdot k}{2}\cdot j^{k-1} + \cdots \\ &\qquad + \frac{(k+1)\cdot k}{2}\cdot j^2 + (k+1)\cdot j + 1\,. \end{aligned}$$

Summing from $j = 1$ to $j = N$ yields

$$\sum_{j=1}^{N} \left\{ (j+1)^{k+1} - j^{k+1} \right\} = (k+1) \cdot S_{k,N} + \frac{(k+1) \cdot k}{2} \cdot S_{k-1,N} + \dots$$
$$+ \frac{(k+1) \cdot k}{2} \cdot S_{2,N} + (k+1) \cdot S_{1,N} + N.$$

The sum on the left collapses to $(N+1)^{k+1} - 1$. We may solve for $S_{k,N}$ and obtain

$$S_{k,N} = \frac{1}{k+1} \cdot \left[(N+1)^{k+1} - 1 - N - \frac{(k+1) \cdot k}{2} \cdot S_{k-1,N} \right.$$
$$\left. - \dots - \frac{(k+1) \cdot k}{2} \cdot S_{2,N} - (k+1) \cdot S_{1,N} \right].$$

We have succeeded in expressing $S_{k,N}$ in terms of $S_{1,N}, S_{2,N}, \dots, S_{k-1,N}$. Thus we may inductively obtain formulas for $S_{k,N}$ for any k. It turns out that

$$\begin{aligned}
S_{1,N} &= 1 + 2 + \cdots + N = \frac{N(N+1)}{2} \\
S_{2,N} &= 1^2 + 2^2 + \cdots + N^2 = \frac{N(N+1)(2N+1)}{6} \\
S_{3,N} &= 1^3 + 2^3 + \cdots + N^3 = \frac{N^2(N+1)^2}{4} \\
S_{4,N} &= 1^4 + 2^4 + \cdots + N^4 = \frac{(N+1)N(2N+1)(3N^2+3N-1)}{30}
\end{aligned}$$

Although the sums treated in this section are not series per se, they are very much in the spirit of our study of series. And they are often useful—for example in computing Riemann sums.

3.5 OPERATIONS ON SERIES

Some operations on series, such as addition, subtraction, and scalar multiplication, are straightforward. Others, such as multiplication, entail subtleties.

3.5.1 Sums and Scalar Products of Series

Proposition 3.5.1. *Let*

$$\sum_{j=1}^{\infty} a_j \quad \textit{and} \quad \sum_{j=1}^{\infty} b_j$$

be convergent series of real or complex numbers that sum to α and β respectively. Then

(a) *The series $\sum_{j=1}^{\infty}(a_j + b_j)$ converges to $\alpha + \beta$.*

(b) *If c is a constant then the series $\sum_{j=1}^{\infty} c \cdot a_j$ converges to $c \cdot \alpha$.*

3.5.2 Products of Series

In order to keep our discussion of multiplication of series as straightforward as possible, we deal at first with absolutely convergent series. It is convenient in this discussion to begin our sum at $j = 0$ instead of $j = 1$. If we wish to multiply

$$\sum_{j=0}^{\infty} a_j \quad \text{and} \quad \sum_{j=0}^{\infty} b_j,$$

then we need to specify what the partial sums of the product series should be. An obvious necessary condition that we wish to impose is that if the first series converges to α and the second converges to β then the product series $\sum_{j=0}^{\infty} c_j$, whatever we define it to be, should converge to $\alpha \cdot \beta$.

The naive method for defining the summands of the product series is to let $c_j = a_j \cdot b_j$. However, a glance at the product of two partial sums of the given series shows that such a definition would be ignoring the distributivity of multiplication over addition.

3.5.3 The Cauchy Product

Cauchy's idea was that the terms for the product series should be

$$c_m \equiv \sum_{j=0}^{m} a_j \cdot b_{m-j}.$$

This particular form for the summands can be easily motivated using power series considerations (which we shall provide later on). For now we concentrate on confirming that this "Cauchy product" of two series really works.

Theorem 3.5.2. *Let $\sum_{j=0}^{\infty} a_j$ and $\sum_{j=0}^{\infty} b_j$ be two absolutely convergent series that converge to limits α and β respectively. Let*

$$c_m = \sum_{j=0}^{m} a_j \cdot b_{m-j} \,.$$

Then the series $\sum_{m=0}^{\infty} c_m$ converges to $\alpha \cdot \beta$.

EXAMPLE 3.5.3. In the Cauchy product of the two conditionally convergent series

$$\sum_{j=0}^{\infty} \frac{(-1)^j}{\sqrt{j+1}} \quad \text{and} \quad \sum_{j=0}^{\infty} \frac{(-1)^j}{\sqrt{j+1}} \,,$$

we see that

$$\begin{aligned} c_m &= \frac{(-1)^0(-1)^m}{\sqrt{1}\sqrt{m+1}} + \frac{(-1)^1(-1)^{m-1}}{\sqrt{2}\sqrt{m}} + \cdots + \frac{(-1)^m(-1)^0}{\sqrt{m+1}\sqrt{1}} \\ &= \sum_{j=0}^{m} (-1)^m \frac{1}{\sqrt{(j+1)\cdot(m+1-j)}}. \end{aligned}$$

Because

$$(j+1)\cdot(m+1-j) \leq (m+1)\cdot(m+1) = (m+1)^2], ,$$

we have

$$|c_m| \geq \sum_{j=0}^{m} \frac{1}{m+1} = 1 \,.$$

The terms of the series $\sum_{m=0}^{\infty} c_m$ therefore do not tend to zero, so the series cannot converge. □

EXAMPLE 3.5.4. The series

$$A = \sum_{j=0}^{\infty} 2^{-j} \qquad \text{and} \qquad B = \sum_{j=0}^{\infty} 3^{-j}$$

are both absolutely convergent. We challenge the reader to calculate the Cauchy product and to verify that that product converges to 3. □

CHAPTER 4

The Topology of the Real Line

4.1 Open and Closed Sets

4.1.1 Open Sets

An *open interval* in $\mathbb{R}$ is any set of the form

$$(a, b) = \{x \in \mathbb{R} : a < x < b\}.$$

A *closed interval* in $\mathbb{R}$ is any set of the form

$$[a, b] = \{x \in \mathbb{R} : a \leq x \leq b\}.$$

See Figure 4.1.

FIGURE 4.1. Open and closed intervals.

The intersection of any two open intervals is either empty (i.e., has no points in it) or is another open interval. The union of two open intervals is either another open interval (if the two component intervals overlap) or else is just two disjoint open intervals.

The key property of an open interval is this:

If I is an open interval and $x \in I$ then there is an $\epsilon > 0$ such that

$$(x - \epsilon, x + \epsilon) \subseteq I.$$

Thus any point in an open interval I has a little interval around it that still lies in I. See Figure 4.2.

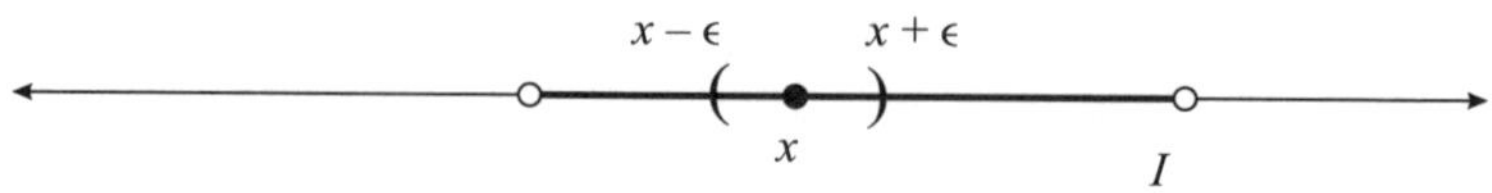

FIGURE 4.2. Neighborhood of a point in an open interval.

We call a set $U \subseteq \mathbb{R}$ *open* if, whenever $x \in U$, there is an $\epsilon > 0$ such that $(x - \epsilon, x + \epsilon) \subseteq U$. Any open set U is the disjoint union of open intervals. See Figure 4.3.

FIGURE 4.3. Structure of an open set.

It may be noted that the union of any number (finite or infinite) of open sets is open. The intersection of finitely many (but not, in general, of infinitely many) open sets is open.

EXAMPLE 4.1.1. Let $U = (3, 4) \cup (7, 9)$. Then U is open. To illustrate this point we take, for instance, the point $x = 8.88 \in U$. Then we may select $\epsilon = 0.1$ and see that $(x - \epsilon, x + \epsilon) = (8.78, 8.98) \subseteq S$. □

4.1.2 CLOSED SETS

A set $E \subseteq \mathbb{R}$ is called *closed* provided that its complement ${}^c E$ is open. Unlike an open set, which is simply a union of intervals, a closed set can be rather complicated (see our discussion of the Cantor set below in Section 4.5). Figure 4.4 depicts a closed set.

FIGURE 4.4. A closed set.

The intersection of any number (finite or infinite) of closed sets is closed. The union of finitely many (but not of infinitely many) closed sets is closed.

EXAMPLE 4.1.2. Let $E = [1, 3] \cup \{5\}$. Then E is closed. To illustrate this point we take $x = 3.15$ in the complement of E. Let $\epsilon = .05$. Then the interval $(x - \epsilon, x + \epsilon) = (3.1, 3.2)$ lies entirely in the complement of E (illustrating that the complement of E is open, hence E is closed). □

4.1.3 CHARACTERIZATION OF OPEN AND CLOSED SETS IN TERMS OF SEQUENCES

Proposition 4.1.3. *Let $S \subseteq \mathbb{R}$ be a set. Then S is closed if and only if each Cauchy sequence $\{s_j\}$ in S has a limit that is also an element of S.*

EXAMPLE 4.1.4. The set $E = [-2, 3] \subseteq \mathbb{R}$ is of course closed. If $\{a_j\}$ is any Cauchy sequence in E then the sequence will have a limit in E. Since the endpoints are included in the set, there is no possibility for the sequence to converge to an exterior point. □

EXAMPLE 4.1.5. Let $S = (-2, 3)$. This set is not closed. The sequence $a_j = -2 + 1/j$ lies in S and has limit -2. Because the limit point -2 does *not* lie in S, the set S fails to be closed. □

It follows from the completeness of the real numbers that any Cauchy sequence has a limit *in* $\mathbb{R}$. The main point of the proposition is that, when the set S is closed, then a Cauchy sequence in S has its limit in S.

Such a characterization cannot hold for open sets. For instance, let $I = (0, 1)$ and let $a_j = 1/(j + 1)$. Then $a_j \in I$ for each j, and the sequence has a limit (namely, the point 0). Yet that limit point is *not* in I.

We may state an auxiliary result which is in fact trivially tautologically equivalent to this last:

Proposition 4.1.6. *Let $U \subseteq \mathbb{R}$ be a set. Then U is open if, whenever $\{a_j\}$ is a sequence in ${}^cU$, then the limit point of the sequence is also in ${}^cU$.*

EXAMPLE 4.1.7. The set $U = (-1, 4) \subseteq \mathbb{R}$ is open. If $\{a_j\}$ is any Cauchy sequence in ${}^cU$ then the sequence has a limit in the complement of U—since the endpoints are included in the complement there is no possibility for the sequence to converge to a point of U. □

4.1.4 FURTHER PROPERTIES OF OPEN AND CLOSED SETS

Let $S \subseteq \mathbb{R}$ be a set. We call $b \in \mathbb{R}$ a *boundary point* of S if every non-trivial neighborhood $(b - \epsilon, b + \epsilon)$ contains both points of S and points of $\mathbb{R} \setminus S$. We denote the set of boundary points of S by ∂S. Refer to Figure 4.5.

A boundary point b might lie in S and might lie in the complement of S. The next example serves to illustrate the concept:

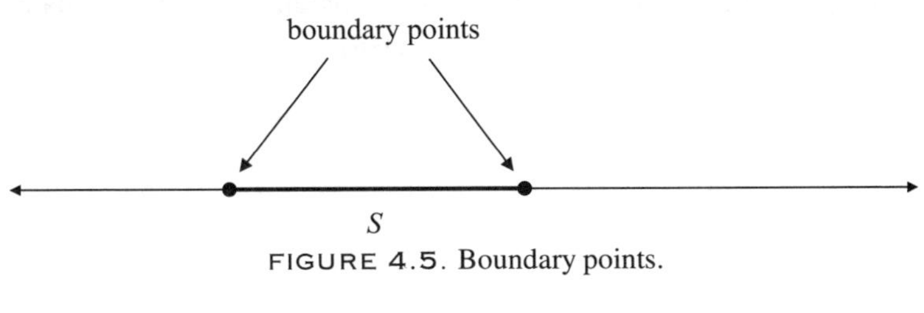

FIGURE 4.5. Boundary points.

EXAMPLE 4.1.8. Let S be the interval $(0, 1)$. Then no point of $(0, 1)$ is in the boundary of S since every point of $(0, 1)$ has a neighborhood that lies inside $(0, 1)$. Also no point of the complement of $[0, 1]$ lies in the boundary of S for a similar reason. The only candidates for elements of the boundary of S are 0 and 1. The point 0 is an element of the boundary since every neighborhood $(0 - \epsilon, 0 + \epsilon)$ contains the points $(0, \epsilon) \subseteq S$ and points $(-\epsilon, 0] \subseteq \mathbb{R} \setminus S$. A similar calculation shows that 1 lies in the boundary of S.

Consider the set $T = [0, 1]$. There are no boundary points in $(0, 1)$, for the same reason as in the first paragraph; and there are no boundary points in $\mathbb{R} \setminus [0, 1]$, since that set is open. Thus the only candidates for elements of the boundary are 0 and 1. As in the first paragraph, they are both boundary points for T.

Neither of the boundary points of S lie in S while both of the boundary points of T lie in T.

For the set $W = [0, 1)$, the points 0, 1 are both boundary points. Clearly 0 lies in W while 1 does not. □

EXAMPLE 4.1.9. The boundary of the set $\mathbb{Q}$ is the entire real line. For if x is any element of $\mathbb{R}$ then every interval $(x - \epsilon, x + \epsilon)$ contains both rational numbers and irrational numbers. □

4.2 OTHER DISTINGUISHED POINTS

4.2.1 INTERIOR POINTS AND ISOLATED POINTS

Definition 4.2.1. Let $S \subseteq \mathbb{R}$. A point $s \in S$ is called an *interior point* of S if there is an $\epsilon > 0$ such that the interval $(s - \epsilon, s + \epsilon)$ lies in S.

A point $t \in S$ is called an *isolated point* of S if there is an $\epsilon > 0$ such that the intersection of the interval $(t - \epsilon, t + \epsilon)$ with S is just the singleton $\{t\}$. See Figure 4.6.

By the definitions given here, an isolated point t of a set $S \subseteq \mathbb{R}$ is a boundary point. For any interval $(t - \epsilon, t + \epsilon)$ contains a point of S (namely t itself) and points of $\mathbb{R} \setminus S$ (since t is isolated).

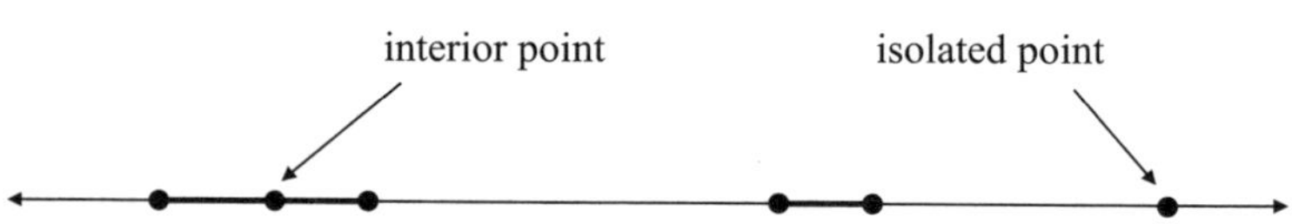

FIGURE 4.6. An isolated point.

A set consisting only of isolated points is called *discrete*. For instance, the integers $\mathbb{Z} \subseteq \mathbb{R}$ is a discrete set. Also the set $\{1, 1/2, 1/3, \ldots\} \subseteq \mathbb{R}$ is discrete.

Proposition 4.2.2. *Let $S \subseteq \mathbb{R}$. Then each point of S is either an interior point or a boundary point.*

EXAMPLE 4.2.3. Let $S = [0, 1]$. Then the interior points of S are the elements of $(0, 1)$. The boundary points of S are the points 0 and 1. The set S has no isolated points.

Let $T = \{1, 1/2, 1/3, \ldots\} \cup \{0\}$. Then the points $1, 1/2, 1/3, \ldots$ are isolated points of T. But 0 is not an isolated point. Every element of T is a boundary point, and there are no others. □

Remark 4.2.4. The interior points of a set S are elements of S, by their very definition; and isolated points of S are elements of S. However, a boundary point of S may or may not be an element of S.

4.2.2 ACCUMULATION POINTS

Definition 4.2.5. Let S be a subset of $\mathbb{R}$. A point x is called an *accumulation point* of S if every neighborhood of x contains infinitely many distinct elements of S. In particular, x is an accumulation point of S if it is the limit of a non-constant sequence in S.

Obviously a closed set contains all its accumulation points.

If x is an accumulation point of S then every open neighborhood of x contains infinitely many elements of S. Hence x is either a boundary point of S or an interior point of S; it cannot be an isolated point of S.

EXAMPLE 4.2.6. Let $S = \{x \in \mathbb{Q} : 0 \leq x \leq 1\}$. Then every point of S is an accumulation point of S. Let $T = \{x \in \mathbb{Z} : 1 \leq x \leq 10\}$. Then no point of T is an accumulation point of T.

Proposition 4.2.7. *Let S be a subset of the real numbers. Then the boundary of S equals the boundary of $\mathbb{R} \setminus S$.*

The next theorem allows us to use the concept of boundary to distinguish open sets from closed sets.

Theorem 4.2.8. *A closed set contains all of its boundary points. An open set contains none of its boundary points.*

EXAMPLE 4.2.9. Let $E = [2, 7] \subseteq \mathbb{R}$. Then E is closed, and E contains its two boundary points $2, 7$.

The set $S = (-4, 0] \subseteq \mathbb{R}$ is not closed, and it is missing one of its boundary points (namely, -4).

The set $U = (-2, 3)$ is open, and it is missing both of its boundary points (-2 and 3).

Proposition 4.2.10. *Every non-isolated boundary point of a set S is an accumulation point of the set S.*

EXAMPLE 4.2.11. Consider the set $S = [-1, 2] \cup \{3\} \cup (5, 7]$. The boundary points of S are $\{-1, 2, 3, 5, 7\}$. The *non-isolated* boundary points are $\{-1, 2, 5, 7\}$. We see that each of these latter is an accumulation point of S.

4.3 BOUNDED SETS

Definition 4.3.1. A subset S of the real numbers is called *bounded* if there is a positive number M such that $|s| \leq M$ for every element s of S.

The next result is one of the great theorems of nineteenth century analysis. It is essentially a restatement of the Bolzano-Weierstrass Theorem of Subsection 2.1.6.

Theorem 4.3.2. *Every bounded, infinite subset of $\mathbb{R}$ has an accumulation point.*

Corollary 4.3.3. *Let $S \subseteq \mathbb{R}$ be a closed and bounded set. If $\{a_j\}$ is any sequence in S then there is a Cauchy subsequence $\{a_{j_k}\}$ that converges to an element of S.*

EXAMPLE 4.3.4. The set $E = [4, 10]$ is a closed and bounded set. Let $\{a_j\}$ be a sequence in E. We may use the *method of bisection* to identify a convergent subsequence.

Write $[4, 10] = [4, 7] \cup [7, 10]$. One of the subintervals will contain infinitely many elements of the sequence. Say that it is $[4, 7]$. Select an element a_{j_1} that lies in $[4, 7]$. Now write $[4, 7] = [4, 5.5] \cup [5.5, 7]$. One of those subintervals will contain infinitely many elements of the sequence. Say that it is $[5.5, 7]$. Select an element a_{j_2}, with $j_2 > j_1$, that lies in $[5.5, 7]$. Continue to bisect and choose, at each stage selecting a subinterval that contains infinitely many elements of the sequence and an element a_{j_k} that is further along in the sequence.

In this manner we obtain the desired subsequence. It is clear that it converges because it lies in a telescoping list of closed intervals that are shrinking to a point (i.e., the limit point). □

4.4 COMPACT SETS

4.4.1 INTRODUCTION

Compact sets are sets (usually infinite) which share many of the most important properties of finite sets. They play an important role in real analysis.

Definition 4.4.1. A set $S \subseteq \mathbb{R}$ is called *compact* if every sequence in S has a subsequence that converges to an element of S.

4.4.2 THE HEINE-BOREL THEOREM

Proposition 4.4.2. *A set is compact if and only if it is closed and bounded.*

In the theory of topology, a different definition of compactness is used. It is equivalent to the one just given. We discuss it here.

Definition 4.4.3. Let S be a subset of the real numbers. A collection of open sets $\{\mathcal{O}_\alpha\}_{\alpha \in A}$ (each $\mathcal{O}_\alpha$ is an open set of real numbers) is called an *open covering* of S if

$$\bigcup_{\alpha \in A} \mathcal{O}_\alpha \supseteq S.$$

EXAMPLE 4.4.4. The collection $\mathcal{C} = \{(1/j, 1)\}_{j=1}^{\infty}$ is an open covering of the interval $I = (0, 1)$. However, no finite subcollection of $\mathcal{C}$ covers I.

The collection $\mathcal{D} = \{(1/j, 1)\}_{j=1}^{\infty} \cup \{(-1/5, 1/5)\} \cup \{(4/5, 6/5\}$ is an open covering of the interval $J = [0, 1]$. However, not all the elements $\mathcal{D}$ are actually needed to cover J. In fact

$$(-1/5, 1/5)\ ,\ (1/6, 1)\ ,\ (4/5, 6/5)$$

cover the interval J—see Figure 4.7. □

FIGURE 4.7. An open covering.

It is the special property displayed in this example that distinguishes compact sets from the point of view of topology. We need another definition:

Definition 4.4.5. If $\mathcal{C}$ is an open covering of a set S and if $\mathcal{D}$ is another open covering of S such that each element of $\mathcal{D}$ is also an element of $\mathcal{C}$ then we call $\mathcal{D}$ a *subcovering* of $\mathcal{C}$.

We call $\mathcal{D}$ a *finite subcovering* if $\mathcal{D}$ has just finitely many elements.

EXAMPLE 4.4.6. The collection of intervals

$$\mathcal{C} = \{(j-1, j+1)\}_{j=1}^{\infty}$$

is an open covering of the set $S = [5, 9]$. The collection

$$\mathcal{D} = \{(j-1, j+1)\}_{j=5}^{\infty}$$

is a subcovering.

The collection

$$\mathcal{E} = \{(4,6), (5,7), (6,8), (7,9), (8,10)\}$$

is a finite subcovering. □

4.4.3 THE TOPOLOGICAL CHARACTERIZATION OF COMPACTNESS

Theorem 4.4.7. *A set $S \subseteq \mathbb{R}$ is compact if and only if every open covering $\mathcal{C} = \{\mathcal{O}_\alpha\}_{\alpha \in A}$ of S has a finite subcovering.*

EXAMPLE 4.4.8. If $A \subseteq B$ and both sets are non-empty then $A \cap B = A \neq \emptyset$. A similar assertion holds when intersecting finitely many non-empty sets $A_1 \supseteq A_2 \supseteq \cdots \supseteq A_k$; then $\cap_{j=1}^{k} A_j = A_k$.

It is possible to have infinitely many non-empty nested sets with null intersection. An example is the sets $I_j = (0, 1/j)$. For all j, we see that $I_j \supseteq I_{j+1}$; yet

$$\bigcap_{j=1}^{\infty} I_j = \emptyset .$$

By contrast, if we take $K_j = [0, 1/j]$ then

$$\bigcap_{j=1}^{\infty} K_j = \{0\}.$$

The next proposition shows that compact sets have the intuitively appealing property of the sets K_j rather than the unsettling and non-intuitive property of the sets I_j. □

Proposition 4.4.9. *Let*

$$K_1 \supseteq K_2 \supseteq \cdots \supseteq K_j \supseteq \cdots$$

be non-empty compact sets of real numbers. Set

$$\mathcal{K} = \bigcap_{j=1}^{\infty} K_j .$$

Then $\mathcal{K}$ is compact and $\mathcal{K} \neq \emptyset$.

4.5 THE CANTOR SET

In this section we describe the construction of a remarkable subset of $\mathbb{R}$ with many pathological properties.

We begin with the unit interval $S_0 = [0, 1]$. We extract from S_0 its open middle third; thus $S_1 = S_0 \setminus (1/3, 2/3)$, which consists of two closed intervals of equal length $1/3$.

We construct S_2 from S_1 by extracting from each of its two intervals the middle third: $S_2 = [0, 1/9] \cup [2/9, 3/9] \cup [6/9, 7/9] \cup [8/9, 1]$. Figure 4.8 shows S_2.

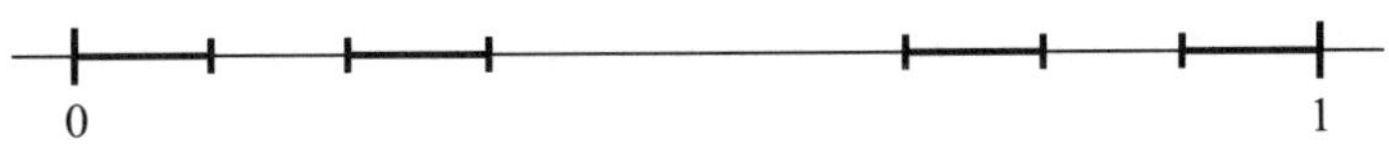

FIGURE 4.8. The set S_2.

Continuing, we construct S_{j+1} from S_j by extracting the open middle third from each of its component subintervals. We define the Cantor set C to be

$$C = \bigcap_{j=1}^{\infty} S_j .$$

Each of the sets S_j is nonempty, closed, and bounded, and hence compact. By Proposition 4.4.9, C is therefore not empty. The set C is closed and bounded, hence compact.

Proposition 4.5.1. *The Cantor set C has zero length, in the sense that* $[0, 1] \setminus C$ *has length* 1.

Idea of the Calculation: In the construction of S_1, we removed from the unit interval one interval of length 3^{-1}. In constructing S_2, we removed two intervals of length 3^{-2}. In constructing S_j, we removed 2^{j-1} intervals of length 3^{-j}. The total length of the intervals removed from the unit interval is

$$\sum_{j=1}^{\infty} 2^{j-1} \cdot 3^{-j} = \frac{1}{3} \sum_{j=0}^{\infty} \left(\frac{2}{3}\right)^j .$$

The total length of the intervals removed is the sum of the geometric series (see Subsection 3.2.3),

$$\frac{1}{3}\left(\frac{1}{1-2/3}\right) = 1\,.$$

Thus the Cantor set has length zero because its complement in the unit interval has length 1. □

Proposition 4.5.2. *The Cantor set is uncountable.*

We can think of each element of the Cantor set as a limit of a sequence of intervals coming from the S_j (see the discussion below). This makes it possible to assign an address (consisting of a sequence of 0's and 1's— at each step we assign 0 for the left interval and 1 for the right interval) to each element of the Cantor set. But there are uncountably many such addresses.

The Cantor set is quite thin (it has zero length) but it is large in the sense that it has uncountably many elements. Also it is compact. The next result reveals a surprising property of this "thin" set:

Theorem 4.5.3. *Let C be the Cantor set and define*

$$S = \{x + y : x \in C, y \in C\}.$$

Then $S = [0, 2]$.

Idea of the Calculation: We sketch the argument.

Since $C \subseteq [0, 1]$ it is clear that $S \subseteq [0, 2]$. For the reverse inclusion, fix an element $t \in [0, 2]$. Our job is to find two elements c and d in C such that $c + d = t$.

First observe that $\{x + y : x \in S_1, y \in S_1\} = [0, 2]$. Therefore there exist $x_1 \in S_1$ and $y_1 \in S_1$ such that $x_1 + y_1 = t$.

Similarly, $\{x + y : x \in S_2, y \in S_2\} = [0, 2]$. Therefore there exist $x_2 \in S_2$ and $y_2 \in S_2$ such that $x_2 + y_2 = t$.

Continuing, we may find for each j numbers x_j and y_j such that $x_j, y_j \in S_j$ and $x_j + y_j = t$. Because $\{x_j\} \subseteq C$ and $\{y_j\} \subseteq C$, there are subsequences $\{x_{j_k}\}$ and $\{y_{j_k}\}$ that converge to real numbers c and d. Since C is compact, we can be sure that $c \in C$ and $d \in C$. But the operation of addition respects limits, thus we may pass to the limit as $k \to \infty$ in the equation

$$x_{j_k} + y_{j_k} = t$$

to obtain

$$c + d = t.$$

Therefore $[0, 2] \subseteq \{x + y : x \in C\}$. This completes the proof. □

Whereas any open set is the union of open intervals, the Cantor set shows us that there is no such structure theorem for closed sets. In fact closed intervals are atypically simple when considered as examples of closed sets.

4.6 CONNECTED AND DISCONNECTED SETS

4.6.1 CONNECTIVITY

Let S be a set of real numbers. We say that S is *disconnected* if it is possible to find a pair of nonempty open sets U and V such that

$$U \cap S \neq \emptyset, V \cap S \neq \emptyset,$$

$$(U \cap S) \cap (V \cap S) = \emptyset,$$

and

$$S = (U \cap S) \cup (V \cap S).$$

If no such U and V exist then we call S *connected*. See Figure 4.9.

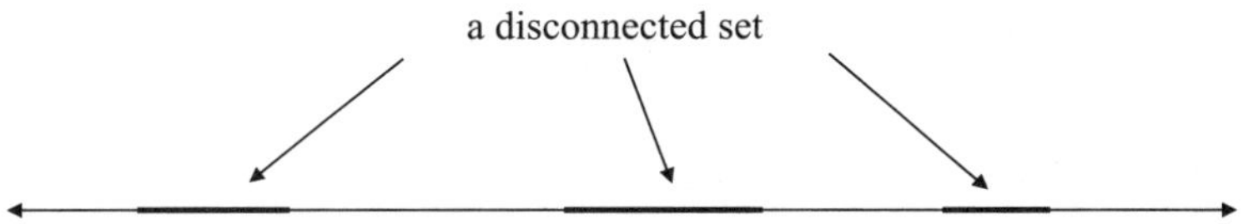

FIGURE 4.9. Connected and disconnected sets.

EXAMPLE 4.6.1. The set $T = \{x \in \mathbb{R} : |x| < 1, x \neq 0\}$ is disconnected. For take $U = \{x : x < 0\}$ and $V = \{x : x > 0\}$. Then

$$U \cap T = \{x : -1 < x < 0\} \neq \emptyset$$

and

$$V \cap T = \{x : 0 < x < 1\} \neq \emptyset.$$

Also $(U \cap T) \cap (V \cap T) = \emptyset$. Clearly $T = (U \cap T) \cup (V \cap T)$, hence T is disconnected. □

EXAMPLE 4.6.2. The set $X = [-1, 1]$ is connected. To see this, suppose to the contrary that there exist open sets U and V such that $U \cap X \neq \emptyset, V \cap X \neq \emptyset, (U \cap X) \cap (V \cap X) = \emptyset$, and

$$S = (U \cap X) \cup (V \cap X).$$

Choose $a \in U \cap X$ and $b \in V \cap X$. Set

$$\alpha = \sup (U \cap [a, b]) .$$

Now $[a, b] \subseteq X$ hence $U \cap [a, b]$ is disjoint from V. Thus $\alpha \le b$. But ${}^cV$ is closed hence $\alpha \notin V$. It follows that $\alpha < b$.

If $\alpha \in U$ then, because U is open, there exists an $\widetilde{\alpha} \in U$ such that $\alpha < \widetilde{\alpha} < b$. This would mean that we chose α incorrectly. Hence $\alpha \notin U$. But $\alpha \notin U$ and $\alpha \notin V$ means $\alpha \notin X$. On the other other hand, α is the supremum of a subset of X (since $a \in X, b \in X$, and X is an interval). Since X is a closed interval, we conclude that $\alpha \in X$. This contradiction shows that X must be connected. □

With small modifications, the discussion in the last example demonstrates that any closed interval is connected. We may similarly see that any open interval or half-open interval is connected. The converse is true as well:

Theorem 4.6.3. *If S is a connected subset of $\mathbb{R}$ then S is an interval.*

The Cantor set is not connected; indeed it is disconnected in a special sense. Call a set S *totally disconnected* if, for each distinct $x \in S$, $y \in S$, there exist disjoint open sets U and V such that $x \in U$, $y \in V$, and $S = (U \cap S) \cup (V \cap S)$.

Proposition 4.6.4. *The Cantor set is totally disconnected.*

4.7 PERFECT SETS

A set $S \subseteq \mathbb{R}$ is called *perfect* if it is non-empty, closed, and if every point of S is an accumulation point of S. The property of being perfect is a rather special one: it means that the set has no isolated points.

Obviously a closed interval $[a, b]$ is perfect. After all, a point x in the interior of the interval is surrounded by an entire open interval $(x - \epsilon, x + \epsilon)$ of elements of the interval; moreover a is the limit of elements from the right and b is the limit of elements from the left.

EXAMPLE 4.7.1. Recall the construction of the Cantor set from the beginning of Section 4.5; in particular, the Cantor set C is the intersction of nested compact sets S_j. *A totally disconnected set*, the Cantor set is perfect.

The Cantor set is certainly closed. Now fix $x \in C$. Then certainly $x \in S_1$. Thus x is in one of the two intervals composing S_1. One (or perhaps both) of the endpoints of that interval does not equal x. Call that endpoint a_1. Likewise $x \in S_2$. Therefore x lies in one of the intervals of S_2. Choose an endpoint a_2 of that interval which does not equal x. Continuing in this fashion, we construct a sequence $\{a_j\}$. *Each of the elements of this sequence lies in the Cantor set* (why?). Finally, $|x - a_j| \leq 3^{-j}$ for each j. Therefore x is the limit of the sequence. We have thus proved that the Cantor set is perfect. □

The fundamental theorem about perfect sets tells us that such a set must be rather large. We have

Theorem 4.7.2. *A non-empty perfect set must be uncountable.*

Corollary 4.7.3. *If $a < b$ then the closed interval $[a, b]$ is uncountable.*

We also have a new way of seeing that the Cantor set is uncountable, since it is perfect:

Corollary 4.7.4. *The Cantor set is uncountable.*

Theorem 4.7.5 (Cantor-Bendixon)**.** *Any uncountable, compact set in $\mathbb{R}$ is the union of a perfect set and a countable set.*

EXAMPLE 4.7.6. Let $E = \{1/j : j = 1, 2, \ldots\} \cup \{-1 - 1/j : j = 1, 2, \ldots\} \cup [-1, 0]$. Then E is compact. Moreover, if we let $A = [-1, 0]$ and $B = \{1/j : j = 1, 2, \ldots\} \cup \{-1 - 1/j : j = 1, 2, \ldots\}$, then A is perfect, B is countable, and $E = A \cup B$. □

CHAPTER 5

Limits and the Continuity of Functions

5.1 Definitions and Basic Properties

5.1.1 Limits

Definition 5.1.1. Let $E \subseteq \mathbb{R}$ be a set and let f be a real-valued function with domain E. Fix a point $P \in \mathbb{R}$ that is either in E or is an accumulation point of E. We say that f has *limit* ℓ at P, and we write

$$\lim_{E \ni x \to P} f(x) = \ell ,$$

with ℓ a real number, if for each $\epsilon > 0$ there is a $\delta > 0$ such that when $x \in E$ and $0 < |x - P| < \delta$ then

$$|f(x) - \ell| < \epsilon .$$

EXAMPLE 5.1.2. Let $E = \mathbb{R} \setminus \{0\}$ and

$$f(x) = x \cdot \sin(1/x) \ \text{ if } \ x \in E .$$

Then $\lim_{x \to 0} f(x) = 0$. To see this, let $\epsilon > 0$. Choose $\delta = \epsilon$. If $0 < |x - 0| < \delta$ then

$$|f(x) - 0| = |x \cdot \sin(1/x)| \le |x| < \delta = \epsilon ,$$

as desired. Thus the limit exists and equals 0. □

5.1.2 A Limit that Does Not Exist

EXAMPLE 5.1.3. Let $E = \mathbb{R}$ and

$$g(x) = \begin{cases} 1 \text{ if } x \text{ is rational} \\ 0 \text{ if } x \text{ is irrational.} \end{cases}$$

(The function g is called the *Dirichlet function.*) Then $\lim_{x \to P} g(x)$ does not exist for any point P of E.

To see this, fix $P \in \mathbb{R}$. Seeking a contradiction, assume that there is a limiting value ℓ for g at P. If this is so then we take $\epsilon = 1/2$ and we can find a $\delta > 0$ such that $0 < |x - P| < \delta$ implies

$$|g(x) - \ell| < \epsilon = \frac{1}{2}. \tag{5.1.3.1}$$

If we take x to be rational then (5.1.3.1) says that

$$|1 - \ell| < \frac{1}{2}, \tag{5.1.3.2}$$

while if we take x irrational then (5.1.3.1) says that

$$|0 - \ell| < \frac{1}{2}. \tag{5.1.3.3}$$

The triangle inequality then gives that

$$\begin{aligned} |1 - 0| &= |(1 - \ell) + (\ell - 0)| \\ &\le |1 - \ell| + |\ell - 0| \\ &< \frac{1}{2} + \frac{1}{2} \\ &= 1\,. \end{aligned}$$

Notice that we have exploited (5.1.3.2) and (5.1.3.3) to obtain the penultimate inequality. This contradiction, that $1 < 1$, allows us to conclude that the limit does not exist at P. □

5.1.3 Uniqueness of Limits

Proposition 5.1.4. *Let f be a function with domain E, and let either $P \in E$ or P be an accumulation point of E. If $\lim_{x \to P} f(x) = \ell$ and $\lim_{x \to P} f(x) = m$ then $\ell = m$.*

The last proposition shows that if a limit is calculated by two different methods, then the same answer will result. While of primarily philosophical interest now, this will be important later.

This is a good time to observe that the limits

$$\lim_{x \to P} f(x) \qquad \text{and} \qquad \lim_{h \to 0} f(P+h)$$

are equal in the sense that if one limit exists then so does the other and they both have the same value. These are two different ways to write the same thing.

5.1.4 PROPERTIES OF LIMITS

In order to facilitate checking that certain limits exist, we now record some elementary properties of the limit. This requires that we first recall how functions are combined.

Suppose that f and g are each functions that have domain E. We define the *sum* or *difference* of f and g to be the function

$$(f \pm g)(x) = f(x) \pm g(x),$$

the *product* of f and g to be the function

$$(f \cdot g)(x) = f(x) \cdot g(x),$$

and the *quotient* of f and g to be

$$\left(\frac{f}{g}\right)(x) = \frac{f(x)}{g(x)}.$$

The quotient is defined only at points x for which $g(x) \neq 0$. See also Section 1.4.

Now we have:

Theorem 5.1.5. *Let f and g be functions with domain E and fix a point P that is either in E or is an accumulation point of E. Assume that*

i) $\lim_{x \to P} f(x) = \ell$,

ii) $\lim_{x \to P} g(x) = m$.

Then

a) $\lim_{x \to P} (f \pm g)(x) = \ell \pm m$,

b) $\lim_{x \to P} (f \cdot g)(x) = \ell \cdot m$,

c) $\lim_{x \to P} (f/g)(x) = \ell/m$ *provided* $m \neq 0$.

EXAMPLE 5.1.6. It is a simple matter to check that if $f(x) = x$ then

$$\lim_{x \to P} f(x) = P$$

for every real P. (For $\epsilon > 0$ we may take $\delta = \epsilon$.) If $g(x) \equiv \alpha$ is the constant function taking value α, then

$$\lim_{x \to P} g(x) = \alpha .$$

It follows from parts **a)** and **b)** of the theorem that if $f(x)$ is any polynomial function then

$$\lim_{x \to P} f(x) = f(P) .$$

Moreover, if $r(x)$ is any *rational function* (quotient of polynomials) then we may also use part **c)** of the theorem to conclude that

$$\lim_{x \to P} r(x) = r(P)$$

for all points P at which the rational function $r(x)$ is defined. □

EXAMPLE 5.1.7. If $0 < x \leq \pi/2$, then $0 < \sin x < x$. This is true because $\sin x$ is the distance from the point $(\cos x, \sin x)$ to the x-axis while x is the distance from that point to the x-axis along an arc. See Figure 5.1. If $\epsilon > 0$ we set $\delta = \epsilon$. If $0 < |x - 0| < \delta$ and $0 < x \leq \pi/2$, then

$$|\sin x - 0| < |x| < \delta = \epsilon .$$

Since $\sin(-x) = -\sin x$, the same result holds when $-\pi/2 \leq x < 0$. Therefore

$$\lim_{x \to 0} \sin x = 0 .$$

Since

$$\cos^2 x = 1 - \sin^2 x ,$$

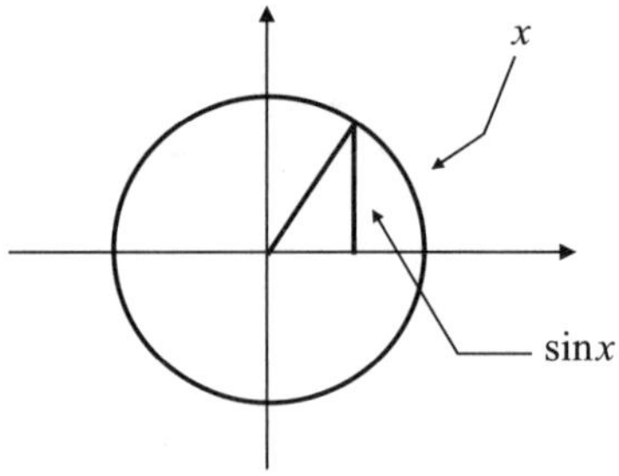

FIGURE 5.1. The function $|\sin x|$ is majorized by $|x|$.

we conclude from the preceding theorem that

$$\lim_{x \to 0} \cos x = 1.$$

For any real number P, we have

$$\begin{aligned} \lim_{x \to P} \sin x &= \lim_{h \to 0} \sin(P + h) \\ &= \lim_{h \to 0} \sin P \cos h + \cos P \sin h \\ &= \sin P \cdot 1 + \cos P \cdot 0 \\ &= \sin P. \end{aligned}$$

We have used parts **a)** and **b)** of the theorem to commute the limit process with addition and multiplication. A similar argument shows that

$$\lim_{x \to P} \cos x = \cos P.$$

□

5.1.5 Characterization of Limits Using Sequences

Proposition 5.1.8. *Let f be a function with domain E and P be either an element of E or an accumulation point of E. Then*

$$\lim_{x \to P} f(x) = \ell$$

if and only if for any sequence $\{a_j\} \subseteq E \setminus \{P\}$ satisfying $\lim_{j \to \infty} a_j = P$ it holds that

$$\lim_{j \to \infty} f(a_j) = \ell.$$

5.2 Continuous Functions

5.2.1 Continuity at a Point

Definition 5.2.1. Let $E \subseteq \mathbb{R}$ be a set and let f be a real-valued function with domain E. Fix a point $P \in E$. We say that f is *continuous* at P if

$$\lim_{x \to P} f(x) = f(P).$$

In the definition of continuity (as distinct from the definition of limit), we require that $P \in E$. This is necessary because we are comparing the value $f(P)$ with the value of the limit.

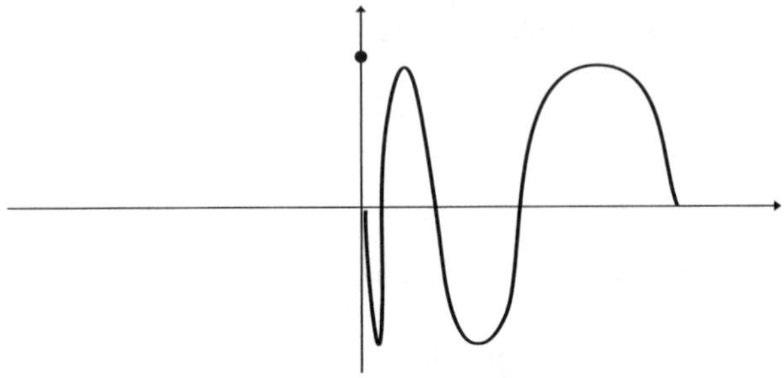

FIGURE 5.2. A discontinuous function.

EXAMPLE 5.2.2. The function

$$h(x) = \begin{cases} \sin 1/x & \text{if } x \neq 0 \\ 1 & \text{if } x = 0 \end{cases}$$

is discontinuous at 0. See Figure 5.2.

The reason is that

$$\lim_{x \to 0} h(x)$$

does not exist. (Details of this assertion are left for you: notice that $h(1/(j\pi)) = 0$ while $h(2/[(4j+1)\pi]) = 1$ for $j = 1, 2, \ldots$.)

The function

$$k(x) = \begin{cases} x \cdot \sin 1/x & \text{if } x \neq 0 \\ 1 & \text{if } x = 0 \end{cases}$$

is also discontinuous at $x = 0$. This time the limit $\lim_{x \to 0} k(x)$ exists (see Example 5.1.2); but the limit does not agree with $k(0)$. Refer to Figure 5.3.

However, the function

$$k(x) = \begin{cases} x \cdot \sin 1/x & \text{if } x \neq 0 \\ 0 & \text{if } x = 0 \end{cases}$$

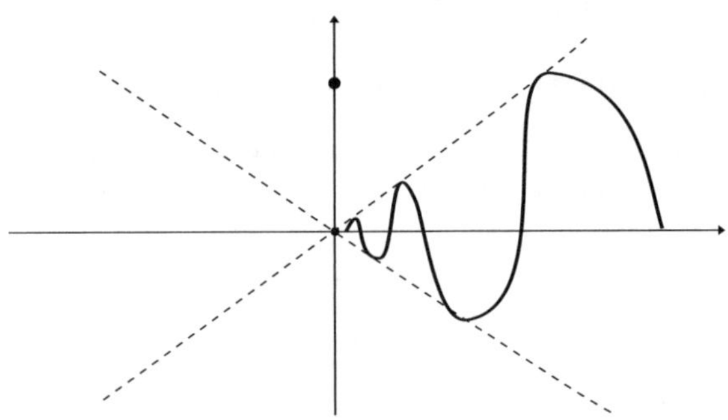

FIGURE 5.3. Another discontinuity.

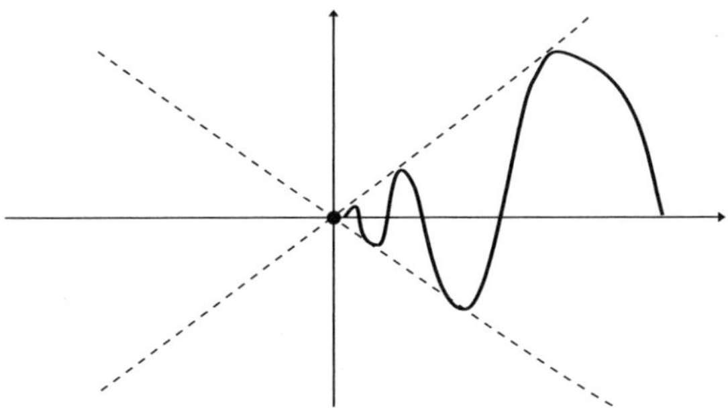

FIGURE 5.4. A continuous function.

is continuous at $x = 0$ because the limit at 0 exists and agrees with the value of the function there. See Figure 5.4.

Theorem 5.2.3. *Let f and g be functions with domain E and let P be a point of E. If f and g are continuous at P then so are $f \pm g$, $f \cdot g$, and (provided $g(P) \neq 0$) f/g.*

Continuous functions may also be characterized using sequences:

Proposition 5.2.4. *Let f be a function with domain E and fix $P \in E$. The function f is continuous at P if and only if, for every sequence $\{a_j\} \subseteq E$ satisfying $\lim_{j\to\infty} a_j = P$, it holds that*

$$\lim_{j\to\infty} f(a_j) = f(P).$$

Proposition 5.2.5. *Let g have domain D and range E and let f have domain E and range H. Let $P \in D$. Assume that g is continuous at P and that f is continuous at $g(P)$. Then $f \circ g$ is continuous at P.*

Remark 5.2.6. It is not the case that if

$$\lim_{x\to P} g(x) = \ell$$

and

$$\lim_{t\to\ell} f(t) = m$$

then

$$\lim_{x\to P} f \circ g(x) = m.$$

A counterexample is given by the functions

$$g(x) \equiv 0$$

$$f(x) = \begin{cases} 2 \text{ if } x \neq 0 \\ 5 \text{ if } x = 0. \end{cases}$$

While $\lim_{x\to 0} g(x) = 0$ and $\lim_{x\to 0} f(x) = 2$, we nevertheless see that $\lim_{x\to 0}(f \circ g)(x) = 5$.

The additional hypothesis that f be continuous at ℓ is necessary to guarantee that the limit of the composition will behave as expected.

5.2.2 THE TOPOLOGICAL APPROACH TO CONTINUITY

Next we explore the topological approach to the concept of continuity. Whereas the analytic approach that we have been discussing so far considers continuity one point at a time, the topological approach considers all points simultaneously. Let us call a function continuous, according to the classical definition that we have been discussing, if it is continuous at every point of its domain.

Definition 5.2.7. Let f be a function with domain E and let $\mathcal{O}$ be any set of real numbers. We define

$$f^{-1}(\mathcal{O}) = \{x \in E : f(x) \in \mathcal{O}\}.$$

We sometimes refer to $f^{-1}(\mathcal{O})$ as the *inverse image* of $\mathcal{O}$ under f.

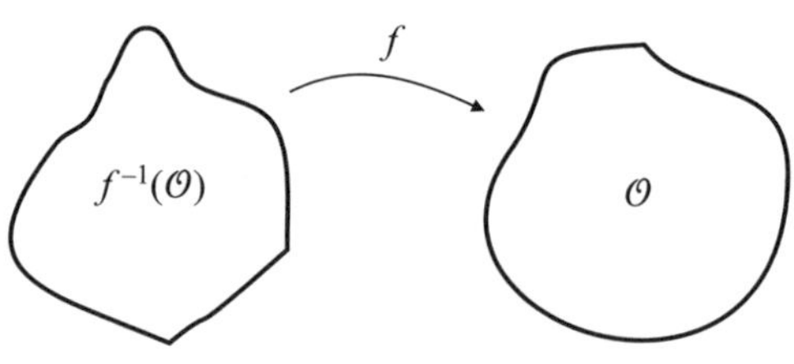

FIGURE 5.5. The inverse image of a set.

Theorem 5.2.8. *Let f be a function with domain E and range F. The function f is continuous (in the classical ϵ-δ sense) if and only if the inverse image under f of any open set in F is the intersection of E with an open set.*

In particular, if E is open then f is continuous if and only if the inverse image of any open set under f is open.

Remark 5.2.9. Since any open subset of the real numbers is a countable union of intervals then, to check that the inverse image under a function f of every open set is open it is enough to check that the inverse image of any open interval is open. This is frequently easy to do, as the next example shows.

EXAMPLE 5.2.10. If $f(x) = x^2$ then the inverse image of an open interval (a, b) is $(-\sqrt{b}, -\sqrt{a}) \cup (\sqrt{a}, \sqrt{b})$ if $a > 0$, is $(-\sqrt{b}, \sqrt{b})$ if $a \leq 0, b \geq 0$, and is $\emptyset$ if $a < b < 0$. Thus the function f is continuous.

By contrast, it is somewhat tedious to give an ϵ-δ proof of the continuity of $f(x) = x^2$. □

EXAMPLE 5.2.11. Let $f : \mathbb{R} \to \mathbb{R}$ be a strictly monotone increasing function. That is to say, $f(a) < f(b)$ whenever $a < b$. Assume that f is continuous. Then it is obvious that f takes an open interval (a, b) to the open interval $(f(a), f(b))$. Likewise, f^{-1} takes an interval (α, β) to the open interval $(f^{-1}(\alpha), f^{-1}(\beta))$. Thus we see immediately that f^{-1} is continuous. It is rather tricky to check the continuity of f^{-1} from the original definitions. □

Corollary 5.2.12. *Let f be a function with domain E. The function f is continuous if and only if the inverse image under f of any closed set F is the intersection of E with some closed set.*

In particular, if E is closed then f is continuous if and only if the inverse image of any closed set under f is closed.

5.3 TOPOLOGICAL PROPERTIES AND CONTINUITY

5.3.1 THE IMAGE OF A FUNCTION

Definition 5.3.1. Let f be a function with domain E and let G be a subset of E. We define

$$f(G) = \{f(x) : x \in G\}.$$

The set $f(G)$ is called the *image* of G under f. See Figure 5.6.

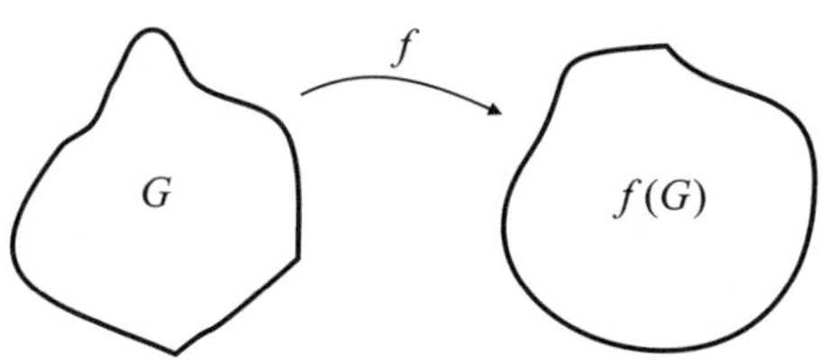

FIGURE 5.6. The image of a set.

Theorem 5.3.2. *The image of a compact set under a continuous function is also compact.*

EXAMPLE 5.3.3. It is not the case that the continuous image of a closed set is closed. For instance, take $f(x) = 1/(1 + x^2)$ and $E = \mathbb{R}$: E is closed and f is continuous but $f(E) = (0, 1]$ is not closed.

It is also not the case that the continuous image of a bounded set is bounded. As an example, take $f(x) = 1/x$ and $E = (0, 1)$. Then E is bounded and f continuous but $f(E) = (1, \infty)$ is unbounded. □

Corollary 5.3.4. *Let f be a function with compact domain K. Then there is a number L such that*

$$|f(x)| \leq L$$

for all $x \in K$.

In fact we can prove an important strengthening of the corollary. Since $f(K)$ is compact, it contains its supremum C and its infimum c. Therefore there must be a number $M \in K$ such that $f(M) = C$ and a number $m \in K$ such that $f(m) = c$. In other words, $f(m) \leq f(x) \leq f(M)$ for all $x \in K$. We summarize:

Theorem 5.3.5. *Let f be a continuous function on a compact set K. Then there exist numbers m and M in K such that $f(m) \leq f(x) \leq f(M)$ for all $x \in K$. We call m an absolute minimum for f on K and M an absolute maximum for f on K. We call $f(m)$ and $f(M)$ the absolute minimum value and absolute maximum value respectively of the function f.*

EXAMPLE 5.3.6. In the last theorem, M and m need not be unique. For instance the function $\sin x$ on the compact interval $[0, 4\pi]$ has an absolute minimum at $3\pi/2$ and $7\pi/2$. It has an absolute maximum at $\pi/2$ and $5\pi/2$. □

5.3.2 UNIFORM CONTINUITY

Now we define a refined type of continuity:

Definition 5.3.7. Let f be a function with domain E. We say that f is *uniformly continuous* on E if, for any $\epsilon > 0$, there is a $\delta > 0$ such that whenever $s, t \in E$ and $|s - t| < \delta$ then $|f(s) - f(t)| < \epsilon$.

The concept of "uniform continuity" differs from "continuity" in that it treats all points of the domain simultaneously: the $\delta > 0$ that is chosen is independent of the points $s, t \in E$. This difference is highlighted by the next example.

EXAMPLE 5.3.8. Consider the function $f(x) = x^2$. Fix a point $P \in \mathbb{R}$, $P > 0$, and let $\epsilon > 0$. In order to guarantee that $|f(x) - f(P)| < \epsilon$ we must have (for $x > 0$)

$$|x^2 - P^2| < \epsilon \qquad \text{or} \qquad |x - P| < \frac{\epsilon}{x + P}.$$

Since x will range over a neighborhood of P, we see that the required δ in the definition of continuity cannot be larger than $\epsilon/(2P)$. In fact the choice $|x - P| < \delta = \epsilon/(2P + 1)$ will do the job.

Thus the choice of δ depends not only on ϵ (which we have come to expect) but also on P. In particular, the function f is not uniformly continuous on $\mathbb{R}$. This is a quantitative reflection of the fact that the graph of f becomes ever steeper as the variable moves to the right (or to the left).

The same calculation shows that the function f, with domain restricted to $[a, b]$, $0 < a < b < \infty$, *is* uniformly continuous. See Figure 5.7. □

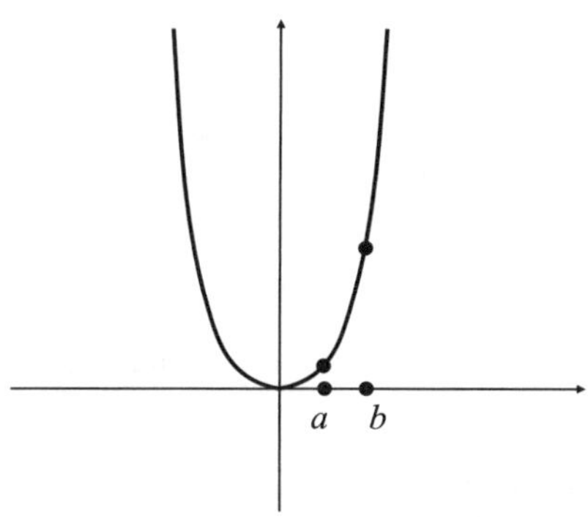

FIGURE 5.7. Uniform continuity.

Now the main result about uniform continuity is the following:

Theorem 5.3.9. *Let f be a continuous function with compact domain K. Then f is uniformly continuous on K.*

EXAMPLE 5.3.10. The function $f(x) = \sin(1/x)$ is continuous on the domain $E = (0, \infty)$ since it is the composition of continuous functions. However, it is not uniformly continuous since

$$\left| f\left(\frac{1}{2j\pi}\right) - f\left(\frac{1}{\frac{(4j+1)\pi}{2}}\right) \right| = 1$$

for $j = 1, 2, \ldots$. Thus, even though the arguments are becoming arbitrarily close together, their images remain bounded apart. We conclude that f cannot be uniformly continuous.

However, if f is considered as a function on any restricted interval of the form $[a, b], 0 < a < b < \infty$, then the preceding theorem tells us that f is uniformly continuous. □

As an exercise, you should check that

$$g(x) = \begin{cases} x \sin(1/x) & \text{if } x \neq 0 \\ 0 & \text{if } x = 0 \end{cases}$$

is uniformly continuous on any interval of the form $[-N, N]$.

Remark 5.3.11. We shall discuss the derivative concept in the next chapter, but we can use it now to learn something about uniform continuity. Let f be a function on the interval $[a, b]$ that has a continuous derivative on that interval. Then the derivative, being a continuous function on a compact interval, is bounded: $|f'(x)| \leq M$ for some M and all $x \in [a, b]$. If $s, t \in [a, b]$ then the mean value theorem tells us that

$$|f(s) - f(t)| = |f'(\xi)| \cdot |s - t|$$

for some point ξ between s and t. This last is bounded by $M \cdot |s - t|$. Thus f is Lipschitz (Subsection 6.3.4), so it is certainly uniformly continuous.

5.3.3 CONTINUITY AND CONNECTEDNESS

Last we note a connection between continuous functions and connectedness.

Theorem 5.3.12. *Let f be a continuous function with domain an open interval I. Suppose that L is a connected subset of I. Then $f(L)$ is connected.*

In other words, the image of an (open or closed) interval under a continuous function is also an interval.

EXAMPLE 5.3.13. Let f be a continuous function on the interval $[a, b]$. Let $\alpha = f(a)$ and $\beta = f(b)$. Choose a number γ that lies between α and β. Is there a number $c \in [a, b]$ such that $f(c) = \gamma$? Because the continuous image of an interval is an interval, the answer is obviously "yes". Thus we have established the important *intermediate value property* for continuous functions. We record this result formally in the next subsection. □

5.3.4 THE INTERMEDIATE VALUE PROPERTY

Corollary 5.3.14. *Let f be a continuous function whose domain contains the interval $[a, b]$. Let γ be a number that lies between $f(a)$ and $f(b)$. Then there is a number c between a and b such that $f(c) = \gamma$.*

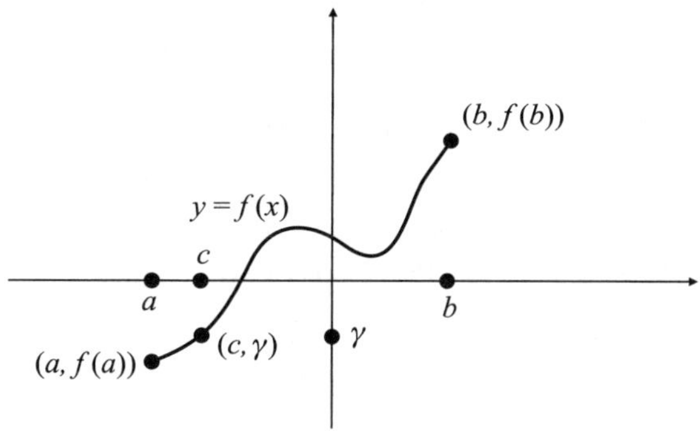

FIGURE 5.8. The intermediate value property.

5.4 MONOTONICITY AND CLASSIFYING DISCONTINUITIES

5.4.1 LEFT AND RIGHT LIMITS

We begin by refining our notion of limit:

Definition 5.4.1. Fix $P \in \mathbb{R}$. Let f be a function with domain E. Fix a point $P \in E$. We say that f has *left limit* ℓ at P, and write

$$\lim_{x \to P-} f(x) = \ell ,$$

if, for every $\epsilon > 0$, there is a $\delta > 0$ such that whenever $P - \delta < x < P$ and $x \in E$ then it holds that

$$|f(x) - \ell| < \epsilon .$$

We say that f has *right limit* m at P, and write

$$\lim_{x \to P+} f(x) = m ,$$

if, for every $\epsilon > 0$, there is a $\delta > 0$ such that whenever $P < x < P + \delta$ and $x \in E$ then it holds that

$$|f(x) - m| < \epsilon .$$

The definitions formalizes the notion of letting x tend to P from the left only or from the right only.

EXAMPLE 5.4.2. Let

$$f(x) = \begin{cases} x^2 & \text{if } 0 \le x < 1 \\ 0 & \text{if } \quad x = 1 \\ 2x - 4 & \text{if } 1 < x < 2 . \end{cases}$$

Then $\lim_{x \to 1^-} f(x) = 1$ while $\lim_{x \to 1^+} f(x) = -2$. The actual *value* of f at 1 is $f(1) = 0$. □

5.4.2 TYPES OF DISCONTINUITIES

Let f be a function with domain E. Let $P \in E$ and assume that f is discontinuous at P. There are two ways in which this discontinuity can occur:

I. If $\lim_{x \to P^-} f(x)$ and $\lim_{x \to P^+} f(x)$ both exist but either do not equal each other or do not equal $f(P)$ then we say that f has a *discontinuity of the first kind* (or sometimes a *simple discontinuity*) at P.

II. If either $\lim_{x \to P^-} f(x)$ does not exist or $\lim_{x \to P^+} f(x)$ does not exist then we say that f has a *discontinuity of the second kind* at P.

See Figure 5.9.

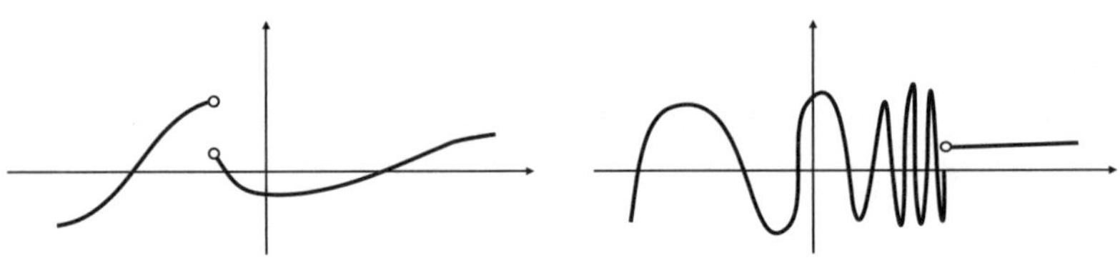

discontinuity of the first kind discontinuity of the second kind

FIGURE 5.9. A discontinuity of the second kind.

EXAMPLE 5.4.3. Define

$$f(x) = \begin{cases} \sin(1/x) & \text{if } x \ne 0 \\ 0 & \text{if } x = 0 \end{cases}$$

$$g(x) = \begin{cases} 1 & \text{if } x > 0 \\ 0 & \text{if } x = 0 \\ -1 & \text{if } x < 0 \end{cases}$$

$$h(x) = \begin{cases} 1 & \text{if } x \text{ is irrational} \\ 0 & \text{if } x \text{ is rational}. \end{cases}$$

Then f has a discontinuity of the second kind at 0 while g has a discontinuity of the first kind at 0. The function h has a discontinuity of the second kind at every point. □

5.4.3 MONOTONIC FUNCTIONS

Definition 5.4.4. Let f be a function whose domain contains an open interval (a, b). We say that f is *monotonically increasing* on (a, b) if, whenever $a < s < t < b$, it holds that $f(s) \le f(t)$. We say that f is *monotonically decreasing* on (a, b) if, whenever $a < s < t < b$, it holds that $f(s) \ge f(t)$. See Figure 5.10.

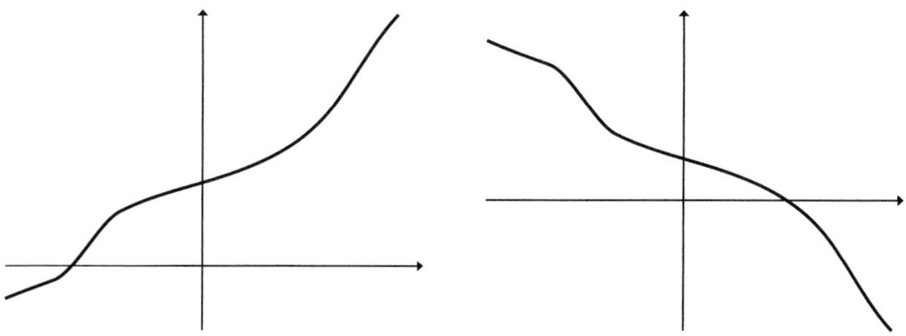

monotonically increasing function monotonically decreasing function

FIGURE 5.10. Monotonicity.

Functions that are either monotonically increasing or monotonically decreasing are simply referred to as "monotonic" or "monotone".

EXAMPLE 5.4.5. The function $f(x) = \sin x$ is monotonically increasing on the interval $[-\pi/2, \pi/2]$ and on all intervals of the form $[(-1 + 4k)\pi/2, (1 + 4k)\pi/2]$. Also the function is monotonically decreasing on the interval $[\pi/2, 3\pi/2]$ and on all intervals of the form $[(1+4k)\pi/2, (3+4k)\pi/2]$. □

As with sequences, the word "monotonic" is superfluous in many contexts. But its use is traditional and occasionally convenient.

Proposition 5.4.6. *Let f be a monotonic function on an open interval (a, b). Then all of the discontinuities of f are of the first kind.*

Corollary 5.4.7. *Let f be a monotonic function on an interval (a, b). Then f has at most countably many discontinuities.*

Theorem 5.4.8. *Let f be a continuous, monotone function whose domain is a compact set K. Let $\mathcal{O}$ be any open set in $\mathbb{R}$. Then $f(K \cap \mathcal{O})$ has the form $f(K) \cap \mathcal{U}$ for some open set $\mathcal{U} \subseteq \mathbb{R}$.*

Suppose that f is a function on (a, b) such that $a < s < t < b$ implies $f(s) < f(t)$. Such a function is called *strictly monotonically increasing*. Likewise, if f is a function on (a, b) such that $a < s < t < b$ implies $f(s) > f(t)$, then we say that f is *strictly monotonically decreasing*. It is clear that a strictly monotonically increasing (resp. strictly monotonically decreasing) function is one-to-one, and hence has an inverse. We summarize:

Theorem 5.4.9. *Let f be a strictly monotone, continuous function with domain $[a, b]$. Then f^{-1} exists and is continuous.*

CHAPTER 6

THE DERIVATIVE

6.1 THE CONCEPT OF DERIVATIVE

6.1.1 THE DEFINITION

Let f be a function with domain an open interval I. If $x \in I$ then the quantity

$$\frac{f(t) - f(x)}{t - x}$$

measures the slope of the chord of the graph of f that connects the points $(x, f(x))$ and $(t, f(t))$. See Figure 6.1. If we let $t \to x$ then the limit of the quantity represented by this "Newton quotient" should represent the slope of the graph *at the point* x. These considerations motivate the definition of the derivative:

Definition 6.1.1. If f is a function with domain an open interval I and if $x \in I$ then the limit

$$\lim_{t \to x} \frac{f(t) - f(x)}{t - x},$$

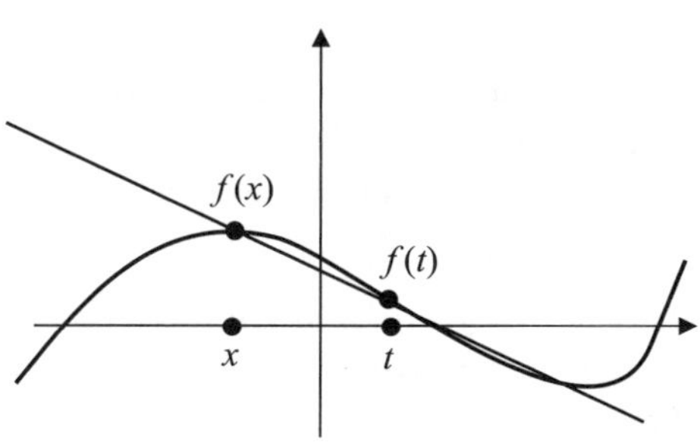

FIGURE 6.1. The derivative.

when it exists, is called the *derivative* of f at x. If the derivative of f at x exists then we say that f is *differentiable* at x. If f is differentiable at every $x \in I$ then we say that f is *differentiable on* I.

We write the derivative of f at x either as

$$f'(x) \quad \text{or} \quad \frac{d}{dx}f \quad \text{or} \quad \frac{df}{dx}.$$

EXAMPLE 6.1.2. Consider the function $f(x) = x^2$ at the point $x = 1$. We endeavor to calculate the derivative:

$$\lim_{t\to x} \frac{f(t) - f(1)}{t - 1} = \lim_{t\to 1} \frac{t^2 - 1^2}{t - 1} = \lim_{t\to 1}(t + 1) = 2\,.$$

Thus the derivative of $f(x) = x^2$ at the point 1 exists and is equal to 2. □

EXAMPLE 6.1.3. Let us calculate the derivative of $g(x) = 1/x$ at an arbitrary point $x \neq 0$. We have

$$\begin{aligned} \lim_{t\to x} \frac{g(t) - g(x)}{t - x} &= \lim_{t\to x} \frac{1/t - 1/x}{t - x} \\ &= \lim_{t\to x} \frac{x - t}{t^2x - tx^2} \\ &= \lim_{t\to x} -\frac{1}{xt} = -\frac{1}{x^2}\,. \end{aligned}$$

We see therefore that the derivative of $g(x) = 1/x$ at an arbitrary point $x \neq 0$ is $g'(x) = -1/x^2$.

6.1.2 PROPERTIES OF THE DERIVATIVE

We begin our discussion of the derivative by establishing some basic properties and relating the notion of derivative to continuity.

Lemma 6.1.4. *If f is differentiable at a point x then f is continuous at x. In particular,* $\lim_{t\to x} f(t) = f(x)$.

Thus all differentiable functions are continuous: differentiability is a stronger property than continuity. It is easy to convince yourself with a picture that if a function f is not continuous at a point then it is certainly not differentiable at that point.

Theorem 6.1.5. *Assume that f and g are functions with domain an open interval I and that f and g are differentiable at $x \in I$. Then $f \pm g$, $f \cdot g$, and f/g are differentiable at x (for f/g we assume that $g(x) \neq 0$). Moreover*

(a) $(f \pm g)'(x) = f'(x) \pm g'(x)$;

(b) $(f \cdot g)'(x) = f'(x) \cdot g(x) + f(x) \cdot g'(x)$;

(c) $\left(\frac{f}{g}\right)'(x) = \frac{g(x) \cdot f'(x) - f(x) \cdot g'(x)}{g^2(x)}$.

EXAMPLE 6.1.6. That $f(x) = x$ is differentiable follows from

$$\lim_{t \to x} \frac{f(t) - f(x)}{t - x} = \lim_{t \to x} \frac{t - x}{t - x} = 1.$$

Hence $f'(x) = 1$ for all x. If $g(x) \equiv c$ is a constant function then

$$\lim_{t \to x} \frac{g(t) - g(x)}{t - x} = \lim_{t \to x} \frac{c - c}{t - x} = 0$$

hence $g'(x) \equiv 0$. It follows now from the theorem that any polynomial function is differentiable.

On the other hand, the function $f(x) = |x|$ is not differentiable at the point $x = 0$. This is so because

$$\lim_{t \to 0^-} \frac{|t| - |0|}{t - x} = \lim_{t \to 0^-} \frac{-t - 0}{t - 0} = -1$$

while

$$\lim_{t \to 0+} \frac{|t| - |0|}{t - x} = \lim_{t \to 0+} \frac{t - 0}{t - 0} = 1.$$

So the required limit does not exist. □

6.1.3 THE WEIERSTRASS NOWHERE DIFFERENTIABLE FUNCTION

Theorem 6.1.7. *Define a function ψ with domain $\mathbb{R}$ by the rule*

$$\psi(x) = \begin{cases} x - n & \text{if } n \leq x < n + 1 \text{ and } n \text{ is even} \\ n + 1 - x & \text{if } n \leq x < n + 1 \text{ and } n \text{ is odd.} \end{cases}$$

The graph of this function is exhibited in Figure 6.2. Then the function

$$f(x) = \sum_{j=1}^{\infty} \left(\frac{3}{4}\right)^j \psi\left(4^j x\right)$$

is continuous at every real x and differentiable at no real x.

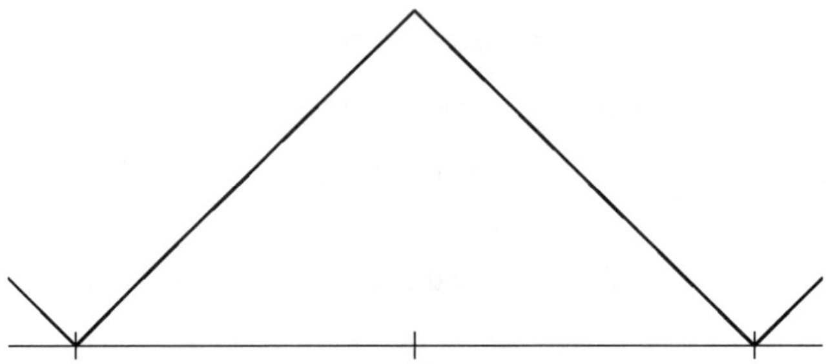

FIGURE 6.2. The function ψ, a component of the Weierstrass nowhere differentiable function.

This startling example of Weierstrass emphasizes the fact that continuity certainly does not imply differentiability.

EXAMPLE 6.1.8. The function

$$g(x) \equiv \sum_{j=1}^{\infty} \left(\frac{3}{16}\right)^{j} \psi\left(4^{j} x\right)$$

has the property that it is continuously differentiable, but not twice differentiable, at any point. The function

$$g_k(x) \equiv \sum_{j=1}^{\infty} \left(\frac{3}{4^{1+k}}\right)^{j} \psi\left(4^{j} x\right)$$

has the property that it is k times continuously differentiable, but not $(k+1)$ times differentiable, at any point. □

6.1.4 THE CHAIN RULE

Next we turn to the Chain Rule.

Theorem 6.1.9. *Let g be a differentiable function on an open interval I and let f be a differentiable function on an open interval that contains the range of g. Then $f \circ g$ is differentiable on the interval I and*

$$(f \circ g)'(x) = f'(g(x)) \cdot g'(x)$$

for each $x \in I$.

Intuitively, if body $\mathcal{F}$ moves f times as fast as body $\mathcal{G}$, and if body $\mathcal{G}$ moves at velocity g, then $\mathcal{F}$ moves at velocity $f \cdot g$.

EXAMPLE 6.1.10. Let $f(x) = x^3$ and $g(x) = \sin x$. Then $(f \circ g)(x) = \sin^3 x$. Thus we have, by the chain rule, that

$$[\sin^3 x]' = f'(g(x)) \cdot g'(x) = 3 \sin^2 x \cdot \cos x . \qquad \square$$

6.2 THE MEAN VALUE THEOREM AND APPLICATIONS

6.2.1 LOCAL MAXIMA AND MINIMA

We begin this section with some remarks about local maxima and minima of functions.

Definition 6.2.1. Let f be a function with domain (a, b). A point $x \in (a, b)$ is called a *local minimum* for f if there is a $\delta > 0$ such that $f(t) \geq f(x)$ for all $t \in (x - \delta, x + \delta)$. A point $x \in (a, b)$ is called a *local maximum* for f if there is a $\delta > 0$ such that $f(t) \leq f(x)$ for all $t \in (x - \delta, x + \delta)$.

Local minima (plural of minimum) and local maxima (plural of maximum) are referred to collectively as *local extrema*.

6.2.2 FERMAT'S TEST

Proposition 6.2.2. *If f is a function with domain (a, b), if f has a local extremum at $x \in (a, b)$, and if f is differentiable at x, then $f'(x) = 0$.*

EXAMPLE 6.2.3. Let

$$f(x) = x + \sin x .$$

Then f is differentiable on the entire real line, $f'(x) = 1 + \cos x$, and f' vanishes at odd multiples of π. Yet, as a glance at the graph of f reveals, f has no local maxima nor minima. This result does not contradict the proposition.

On the other hand, let

$$g(x) = \sin x .$$

Then g has local (indeed global) maxima at points of the form $x = (4k + 1)\pi/2$, and g' vanishes at those points as well. Also g has local (indeed global) minima at points of the form $x = (4k + 3)\pi/2$, and g' vanishes at those points. These results about the function g confirm the proposition. $\square$

6.2.3 DARBOUX'S THEOREM

Before going on to mean value theorems, we provide a striking application of the proposition:

Theorem 6.2.4. *Let f be a differentiable function on an open interval I. Pick points $s < t$ in I and suppose that $f'(s) < \rho < f'(t)$. Then there is a point u between s and t such that $f'(u) = \rho$.*

If f' were a continuous function then the theorem would just be a special instance of the intermediate value property of continuous functions (see Corollary 5.3.14). But derivatives need not be continuous.

EXAMPLE 6.2.5. Consider the function

$$f(x) = \begin{cases} x^2 \cdot \sin(1/x) & \text{if } x \neq 0 \\ 0 & \text{if } x = 0\,. \end{cases}$$

Verify for yourself that $f'(0)$ exists and vanishes but $\lim_{x\to 0} f'(x)$ does not exist. So f' is not continuous at 0. □

This example illustrates the significance of the theorem. Since f' will always satisfy the intermediate value property (even when it is not continuous), its discontinuities cannot be of the first kind. In other words:

> If f is a differentiable function on an open interval I then the discontinuities of f' are all of the second kind.

6.2.4 THE MEAN VALUE THEOREM

Next we turn to the simplest form of the mean value theorem, known as Rolle's theorem.

Theorem 6.2.6 (Rolle). *Let f be a continuous function on the closed interval $[a,b]$ that is differentiable on (a,b). If $f(a) = f(b) = 0$ then there is a point $\xi \in (a,b)$ such that $f'(\xi) = 0$. See Figure 6.3.*

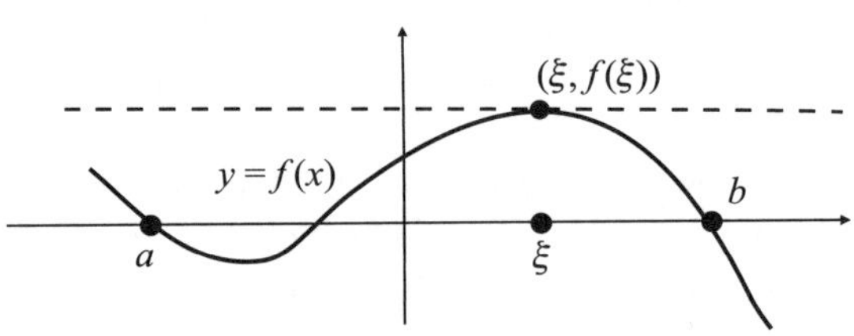

FIGURE 6.3. Rolle's theorem.

EXAMPLE 6.2.7. Let

$$h(x) = xe^x \sin x + \sin 2x \,.$$

Then h satisfies the hypotheses of Rolle's theorem with $a = 0$ and $b = \pi$. We can be sure, therefore, that there is a point ξ between 0 and π so that $h'(\xi) = 0$, even though it may be rather difficult to say exactly what that point is. □

EXAMPLE 6.2.8. The point ξ in Rolle's Theorem need not be unique. If $f(x) = x^3 - x^2 - 2x$ on the interval $[-1, 2]$ then $f(-1) = f(2) = 0$ and f' vanishes at *two* points of the interval $(-1, 2)$. □

If you rotate the graph of a function satisfying the hypotheses of Rolle's Theorem, the result suggests that for any continuous function f on an interval $[a, b]$, differentiable on (a, b), we should be able to relate the slope of the chord connecting $(a, f(a))$ and $(b, f(b))$ with the value of f' at some interior point. That is the content of the mean value theorem:

Theorem 6.2.9. *Let f be a continuous function on the closed interval $[a, b]$ that is differentiable on (a, b). There exists a point $\xi \in (a, b)$ such that*

$$\frac{f(b) - f(a)}{b - a} = f'(\xi).$$

See Figure 6.4.

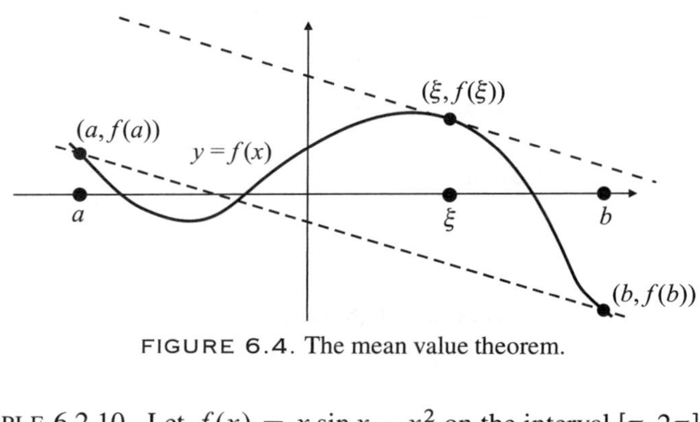

FIGURE 6.4. The mean value theorem.

EXAMPLE 6.2.10. Let $f(x) = x \sin x - x^2$ on the interval $[\pi, 2\pi]$. Since $f(\pi) = -\pi^2$ and $f(2\pi) = -4\pi^2$, we see that

$$\frac{f(2\pi) - f(\pi)}{2\pi - \pi} = -3\pi \,.$$

The mean value theorem guarantees that there is a point ξ between π and 2π at which the derivative of f equals -3π. It would be difficult to say concretely where that point is. □

Corollary 6.2.11. *If f is a differentiable function on the open interval I and if $f'(x) = 0$ for all $x \in I$ then f is a constant function.*

This is an immediate application of the mean value theorem.

Corollary 6.2.12. *If f is differentiable on an open interval I and $f'(x) \geq 0$ for all $x \in I$ then f is monotone increasing on I; that is, if $s < t$ are elements of I then $f(s) \leq f(t)$.*

If f is differentiable on an open interval I and $f'(x) \leq 0$ for all $x \in I$ then f is monotone decreasing on I; that is, if $s < t$ are elements of I then $f(s) \geq f(t)$.

Again the mean value theorem gives the result.

6.2.5 Examples of the Mean Value Theorem

EXAMPLE 6.2.13. Let us verify that

$$\lim_{x \to +\infty} \left(\sqrt{x+5} - \sqrt{x} \right) = 0.$$

Here the limit operation means that, for any $\epsilon > 0$, there is an $N > 0$ such that $x > N$ implies that the expression in parentheses has absolute value less than ϵ.

Define $f(x) = \sqrt{x}$ for $x > 0$. Then the expression in parentheses is just $f(x+5) - f(x)$. By the Mean Value Theorem this equals

$$f'(\xi) \cdot 5$$

for some $x < \xi < x + 5$. But this last expression is

$$\frac{1}{2} \cdot \xi^{-1/2} \cdot 5.$$

By the bounds on ξ, this is bounded above by

$$\frac{5}{2} x^{-1/2}.$$

Clearly, as $x \to +\infty$, this expression tends to 0. □

A powerful tool in analysis is a generalization of the usual Mean Value Theorem that is due to A. L. Cauchy (1789–1857):

Theorem 6.2.14. *Let f and g be continuous functions on the interval $[a, b]$ that are both differentiable on the interval (a, b). Then there is a point $\xi \in (a, b)$ such that*

$$\frac{f(b) - f(a)}{g(b) - g(a)} = \frac{f'(\xi)}{g'(\xi)}.$$

The usual Mean Value Theorem can be obtained from Cauchy's by taking $g(x)$ to be the function x. We conclude this section by illustrating a typical application of the result.

EXAMPLE 6.2.15. Let f be a differentiable function on an interval I such that f' is differentiable at a point $x \in I$. Then

$$\lim_{h \to 0+} \frac{2(f(x+h) + f(x-h) - 2f(x))}{h^2} = (f')'(x) = f''(x).$$

To see this, fix x and define $\mathcal{F}(h) = f(x+h) + f(x-h) - 2f(x)$ and $\mathcal{G}(h) = h^2$. Then

$$\frac{2(f(x+h) + f(x-h) - 2f(x))}{h^2} = \frac{\mathcal{F}(h) - \mathcal{F}(0)}{\mathcal{G}(h) - \mathcal{G}(0)}.$$

According to Cauchy's Mean Value Theorem, there is a ξ between 0 and h such that the last line equals

$$\frac{\mathcal{F}'(\xi)}{\mathcal{G}'(\xi)}.$$

Writing this expression out gives

$$\frac{f'(x+\xi) - f'(x-\xi)}{2\xi} = \frac{1}{2} \cdot \frac{f'(x+\xi) - f'(x)}{\xi} + \frac{1}{2} \cdot \frac{f'(x-\xi) - f'(x)}{-\xi},$$

and the last line tends, by the definition of the derivative, to the quantity $(f')'(x) = f''(x)$. □

Cauchy's mean value theorem is also useful in proving l'Hôpital's rule (see below).

6.3 FURTHER RESULTS ON THE THEORY OF DIFFERENTIATION

6.3.1 L'HÔPITAL'S RULE

l'Hôpital's Rule (actually due to his teacher J. Bernoulli (1667-1748)) is a useful device for calculating limits, and a nice application of the Cauchy Mean Value Theorem. Here we present a special case of the theorem.

Theorem 6.3.1 (l'Hôpital). *Suppose that f and g are differentiable functions on an open interval I and that $p \in I$. If*

$$\lim_{x \to p} f(x) = \lim_{x \to p} g(x) = 0$$

and if

$$\lim_{x \to p} \frac{f'(x)}{g'(x)} \tag{6.3.1.1}$$

exists and equals a real number ℓ then

$$\lim_{x \to p} \frac{f(x)}{g(x)} = \ell.$$

Theorem 6.3.2 (l'Hôpital). *Suppose that f and g are differentiable functions on an open interval I and that $p \in I$. If*

$$\lim_{x \to p} f(x) = \lim_{x \to p} g(x) = \pm\infty$$

and if

$$\lim_{x \to p} \frac{f'(x)}{g'(x)} \tag{6.3.2.1}$$

exists and equals a real number ℓ then

$$\lim_{x \to p} \frac{f(x)}{g(x)} = \ell.$$

EXAMPLE 6.3.3. Let us calculate

$$\lim_{x \to 1} \frac{x-1}{\ln x}.$$

We see that the hypotheses of l'Hôpital's first rule are satisfied. Call the desired limit ℓ. Then

$$\ell = \lim_{x \to 1} \frac{x-1}{\ln x} = \lim_{x \to 1} \frac{(x-1)'}{(\ln x)'} = \lim_{x \to 1} \frac{1}{1/x} = 1.$$

Thus the limit we seek to calculate equals 1. □

EXAMPLE 6.3.4. To calculate the limit

$$\lim_{x \to 0} x^x ,$$

we set

$$A = \ln [x^x] = x \ln x = \frac{\ln x}{1/x}$$

and notice that $\lim_{x \to 0} A$ satisfies the hypotheses of the second version of l'Hôpital's rule. Applying l'Hôpital, we find that the limit of A is 0, hence the original limit is 1. □

6.3.2 DERIVATIVE OF AN INVERSE FUNCTION

Proposition 6.3.5. *Let f be an invertible function on an interval (a, b) with non-zero derivative at a point $x \in (a, b)$. Let $X = f(x)$. Then $\left(f^{-1}\right)'(X)$ exists and equals $1/f'(x)$. See Figure 6.5.*

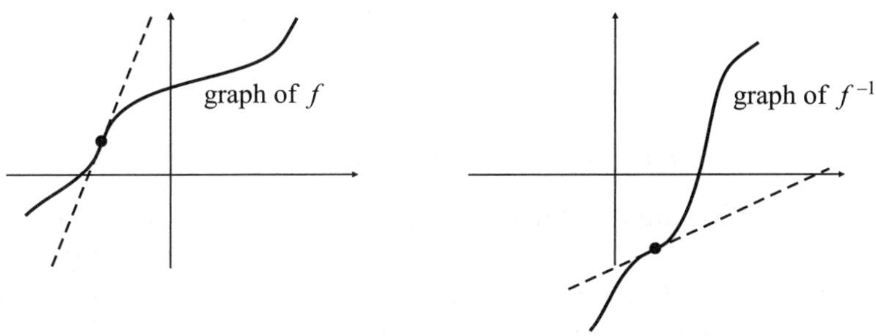

FIGURE 6.5. The derivative of the inverse function.

EXAMPLE 6.3.6. We know that the function $f(x) = x^k, k$ a positive integer, is one-to-one and differentiable on the interval (0, 1). Moreover the derivative $k \cdot x^{k-1}$ never vanishes on that interval. Therefore the proposition applies and we find for $X \in (0, 1) = f((0, 1))$ that

$$\begin{aligned} \left(f^{-1}\right)'(X) &= \frac{1}{f'(x)} = \frac{1}{f'(X^{1/k})} \\ &= \frac{1}{k \cdot X^{1-1/k}} = \frac{1}{k} \cdot X^{\frac{1}{k}-1}. \end{aligned}$$

In other words,

$$\left(X^{1/k}\right)' = \frac{1}{k} \cdot X^{\frac{1}{k}-1}.$$

□

6.3.3 HIGHER DERIVATIVES

If f is a differentiable function on an open interval I then we may ask whether the function f' is differentiable. If it is, we denote its derivative by

$$f'' \text{ or } f^{(2)} \text{ or } \frac{d^2}{dx^2}f \text{ or } \frac{d^2 f}{dx^2},$$

and call it the *second derivative* of f. Likewise the derivative of the $(k-1)^{\text{st}}$ derivative, if it exists, is called the kth derivative and is denoted

$$f^{''\cdots'} \text{ or } f^{(k)} \text{ or } \frac{d^k}{dx^k}f \text{ or } \frac{d^k f}{dx^k}.$$

We cannot even consider whether $f^{(k)}$ exists at a point unless $f^{(k-1)}$ exists in a *neighborhood* of that point.

EXAMPLE 6.3.7. Let $f(x) = x^2 \ln x$. Then

$$f'(x) = 2x \ln x + x\,, \quad f''(x) = 2 \ln x + 3\,, \quad f'''(x) = \frac{2}{x}\,.$$

□

6.3.4 CONTINUOUS DIFFERENTIABILITY

If f is k times differentiable on an open interval I and if each of the derivatives $f^{(1)}, f^{(2)}, \ldots, f^{(k)}$ is continuous on I then we say that f is *k times continuously differentiable* on I. Obviously there is some redundancy in this definition since the continuity of $f^{(j-1)}$ follows from the existence of $f^{(j)}$. Thus only the continuity of the last derivative $f^{(k)}$ need be checked. Continuously differentiable functions are useful tools in analysis. We denote the class of k-times continuously differentiable functions on I by $C^k(I)$.

EXAMPLE 6.3.8. For $k = 1, 2, \ldots$ the function

$$f_k(x) = \begin{cases} x^{k+1} & \text{if } x \geq 0 \\ -x^{k+1} & \text{if } x < 0 \end{cases}$$

will be k times continuously differentiable on $\mathbb{R}$ but will fail to be $k+1$ times differentiable at $x = 0$. More dramatically, an analysis similar to the one we used on the Weierstrass nowhere differentiable function shows that the function

$$g_k(x) = \sum_{j=1}^{\infty} \frac{3^j}{4^{j+jk}} \sin(4^j x)$$

is k times continuously differentiable on $\mathbb{R}$ but will not be $k+1$ times differentiable at any point (this function, with $k = 0$, was the original Weierstrass example). □

A more refined notion of smoothness of functions is that of Lipschitz or Hölder continuity. If f is a function on an open interval I and $0 < \alpha \leq 1$ then we say that f satisfies a *Lipschitz condition of order* α on I if there is a constant M such that, for all $s, t \in I$, we have

$$|f(s) - f(t)| \leq M \cdot |s - t|^\alpha.$$

Such a function is said to be of class $\text{Lip}_\alpha(I)$. Clearly a function of class Lip_α is uniformly continuous on I. For if $\epsilon > 0$ then we may take $\delta = (\epsilon/M)^{1/\alpha}$: then, for $|s - t| < \delta$, we have

$$|f(s) - f(t)| \leq M \cdot |s - t|^\alpha < M \cdot \epsilon/M = \epsilon.$$

EXAMPLE 6.3.9. Let $f(x) = x^2$. Then f is not in Lip_1 on the entire real line. For

$$\left|\frac{f(s) - f(t)}{s - t}\right| = |s + t|,$$

which grows without bound when s, t are large and positive.

But f is Lip_1 on any bounded interval $[a, b]$. For, if $s, t \in [a, b]$, then

$$\left|\frac{f(s) - f(t)}{s - t}\right| = |s + t| \leq 2(|a| + |b|).$$ □

EXAMPLE 6.3.10. When $\alpha > 1$, the class Lip_α contains only constant functions. For in this instance the inequality

$$|f(s) - f(t)| \leq M \cdot |s - t|^\alpha$$

entails

$$\left|\frac{f(s) - f(t)}{s - t}\right| \leq M \cdot |s - t|^{\alpha - 1}.$$

Because $\alpha - 1 > 0$, letting $s \to t$ yields that $f'(t)$ exists for every $t \in I$ and equals 0. It follows from Corollary 6.2.11 of the last section that f is constant on I. □

Instead of trying to extend the given definition of $\text{Lip}_\alpha(I)$ to $\alpha > 1$ it is customary to define classes of functions $C^{k,\alpha}$, for $k = 0, 1, \ldots$ and $0 < \alpha \leq 1$, by the condition that f be of class C^k on I and that $f^{(k)}$

be an element of $\mathrm{Lip}_\alpha(I)$. We leave it as an exercise for you to verify that $C^{k,\alpha} \subseteq C^{\ell,\beta}$ if either $k > \ell$ or both $k = \ell$ and $\alpha \geq \beta$.

In more advanced studies in analysis, it is appropriate to replace $\mathrm{Lip}_1(I)$, and more generally $C^{k,1}$, with another space (invented by Antoni Zygmund, 1900–1992) defined in a more subtle fashion. In fact it uses the expression $|f(x+h)+f(x-h)-2f(x)|$ that we saw earlier in Example 6.2.15. See [KRA3] for further details on these matters.

CHAPTER 7

THE INTEGRAL

7.1 THE CONCEPT OF INTEGRAL

7.1.1 PARTITIONS

The integral is a generalization of the summation process. That is the point of view that we shall take in this chapter.

Definition 7.1.1. Let $[a, b]$ be a closed interval in $\mathbb{R}$. A finite, ordered set of points $\mathcal{P} = \{x_0, x_1, x_2, \ldots, x_{k-1}, x_k\}$ such that

$$a = x_0 \leq x_1 \leq x_2 \leq \cdots \leq x_{k-1} \leq x_k = b$$

is called a *partition* of $[a, b]$. Refer to Figure 7.1.

If $\mathcal{P}$ is a partition of $[a, b]$, then we let I_j denote the interval $[x_{j-1}, x_j]$, $j = 1, 2, \ldots, k$. The symbol Δ_j denotes the *length* of I_j. The *mesh* of $\mathcal{P}$, denoted by $m(\mathcal{P})$, is defined to be $\max_j \Delta_j$.

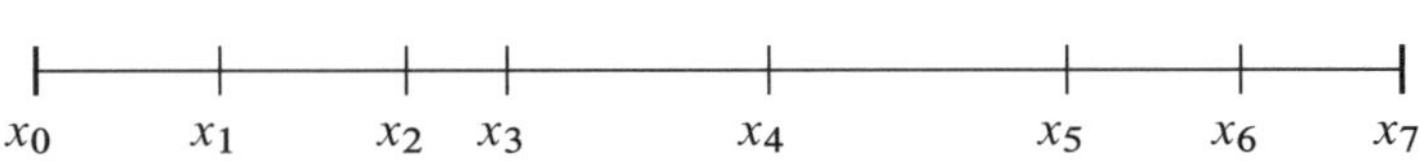

FIGURE 7.1. A partition.

The points of a partition need not be equally spaced, nor must they be distinct from each other.

EXAMPLE 7.1.2. The set $\mathcal{P} = \{0, 1, 1, 9/8, 2, 5, 21/4, 23/4, 6\}$ is a partition of the interval $[0, 6]$ with mesh 3 (because $I_5 = [2, 5]$, with length 3, is the longest interval in the partition). □

Definition 7.1.3. Let $[a, b]$ be an interval and let f be a function with domain $[a, b]$. If $\mathcal{P} = \{x_0, x_1, x_2, \ldots, x_{k-1}, x_k\}$ is a partition of $[a, b]$ and if,

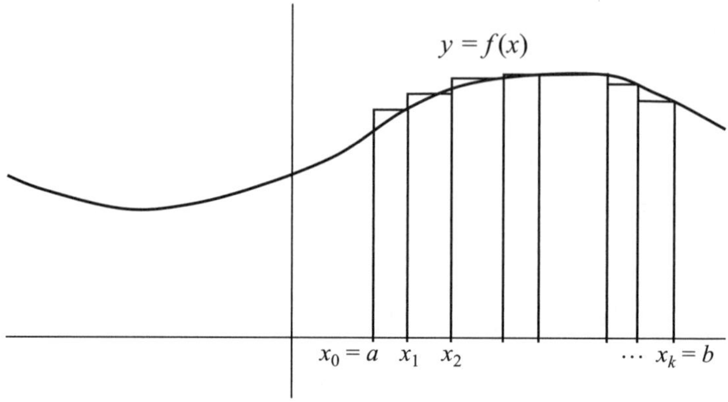

FIGURE 7.2. A Riemann sum.

for each j, s_j is an (arbitrarily chosen) element of $I_j = [x_{j-1}, x_j]$, then the corresponding *Riemann sum* is defined to be

$$\mathcal{R}(f, \mathcal{P}) = \sum_{j=1}^{k} f(s_j)\,\Delta_j .$$

See Figure 7.2.

Remark 7.1.4. In many applications, it is useful to choose s_j to be the right endpoint (or the left endpoint) of the interval I_j. In a theoretical development, it is most convenient to leave the s_j unspecified.

EXAMPLE 7.1.5. Let $f(x) = x^2 - x$ and $[a, b] = [1, 4]$. Define the partition $\mathcal{P} = \{1, 3/2, 2, 7/3, 4\}$ of this interval. We select points $s_1 = 1$, $s_2 = 7/4$, $s_3 = 7/3$, $s_4 = 3$. Then a Riemann sum for this f and $\mathcal{P}$ is

$$\begin{aligned}\mathcal{R}(f, \mathcal{P}) &= \left(1^2 - 1\right) \cdot \frac{1}{2} + \left((7/4)^2 - (7/4)\right) \cdot \frac{1}{2} \\ &\quad + \left((7/3)^2 - (7/3)\right) \cdot \frac{1}{3} + \left(3^2 - 3\right) \cdot \frac{5}{3} \\ &= \frac{10103}{864}.\end{aligned}$$

□

Remark 7.1.6. We stress that the Riemann sum constructed in this last example is not the only one possible. Another, equally valid, Riemann sum would be

$$\begin{aligned}\mathcal{R}(f,\mathcal{P}) &= \left((3/2)^2 - 3/2\right)\cdot\frac{1}{2} + \left(2^2-2\right)\cdot\frac{1}{2}\\ &\quad + \left((7/3)^2 - (7/3)\right)\cdot\frac{1}{3} + \left(4^2-4\right)\cdot\frac{5}{3}.\end{aligned}$$

This time we have chosen $s_1 = 3/2$, $s_2 = 2$, $s_3 = 7/3$, $s_4 = 4$. There is considerable latitude in the choice of the s_j.

Definition 7.1.7. Let $[a,b]$ be an interval and f a function with domain $[a,b]$. We say that *the Riemann sums of f tend to a limit ℓ as $m(\mathcal{P})$ tends to* 0 if, for any $\epsilon > 0$, there is a $\delta > 0$ such that if $\mathcal{P}$ is any partition of $[a,b]$ with $m(\mathcal{P}) < \delta$ then $|\mathcal{R}(f,\mathcal{P}) - \ell| < \epsilon$ for every choice of $s_j \in I_j$ (i.e., for every possible choice of Riemann sum with mesh less than δ).

Definition 7.1.8. A function f on a closed interval $[a,b]$ is said to be *Riemann integrable* on $[a,b]$ if the Riemann sums of $\mathcal{R}(f,\mathcal{P})$ tend to a finite limit as $m(\mathcal{P})$ tends to zero.

The value of the limit, when it exists, is called the *Riemann integral* of f over $[a,b]$ and is denoted by

$$\int_a^b f(x)\,dx.$$

EXAMPLE 7.1.9. Let $f(x) = x^2$. For N a positive integer consider the partition $\mathcal{P} = \{0, 1/N, 2/N, \ldots, (N-1)/N, 1\}$ of the interval $[0,1]$. To keep the discussion simple, we will choose the point s_j to be the right endpoint of the interval $[(j-1)/N, j/N]$ for each j (it turns out that, for a continuous function f, this results in no loss of generality). The corresponding Riemann sum is

$$\mathcal{R}(f,\mathcal{P}) = \sum_{j=1}^{N}\left(\frac{j}{N}\right)^2\cdot\frac{1}{N}\cdot = \frac{1}{N^3}\sum_{j=1}^{N} j^2.$$

Now we may use the formula that we discussed at the end of Section 3.4 to see that this last equals

$$\frac{1}{N^3}\cdot\frac{N(N+1)(2N+1)}{6}.$$

As $N \to \infty$, this expression tends to $1/3$. We conclude that

$$\int_0^1 x^2\,dx = \frac{1}{3}.$$

□

7.1.2 Refinements of Partitions

The basic idea in the theory of the Riemann integral is that refining a partition makes the Riemann sum more closely approximate the desired integral.

Remark 7.1.10. We mention now a useful fact that will be formalized in later remarks. Suppose that f is Riemann integrable on $[a, b]$ with the value of the integral being ℓ. Let $\epsilon > 0$. Then, as stated in the definition (with $\epsilon/3$ replacing ϵ), there is a $\delta > 0$ such that if $\mathcal{Q}$ is a partition of $[a, b]$ of mesh smaller than δ then $|\mathcal{R}(f, \mathcal{Q}) - \ell| < \epsilon/3$. It follows that, if $\mathcal{P}$ and $\mathcal{P}'$ are partitions of $[a, b]$ of mesh smaller than δ, then

$$\left|\mathcal{R}(f, \mathcal{P}) - \mathcal{R}(f, \mathcal{P}')\right| \le |\mathcal{R}(f, \mathcal{P}) - \ell| + |\ell - \mathcal{R}(f, \mathcal{P}')| < \frac{\epsilon}{3} + \frac{\epsilon}{3} = \frac{2\epsilon}{3}.$$

Note, however, that we may choose $\mathcal{P}'$ to equal the partition $\mathcal{P}$. Also we may for each j choose the point s_j, where f is evaluated for the Riemann sum over $\mathcal{P}$, to be the point where f very nearly assumes its supremum on I_j. Then we may for each j choose the point s'_j, where f is evaluated for the Riemann sum over $\mathcal{P}'$, to be a point where f very nearly assumes its infimum on I_j. It easily follows that when the mesh of $\mathcal{P}$ is less than δ then

$$\sum_j \left(\sup_{I_j} f - \inf_{I_j} f \right) \Delta_j < \epsilon. \tag{7.1.10.1}$$

Inequality (7.1.10.1) is a sort of Cauchy condition for the integral. This consequence of integrability will prove useful to us in some of the discussions in this and the next section.

Definition 7.1.11. If $\mathcal{P}$ and $\mathcal{Q}$ are partitions of an interval $[a, b]$ then we say that $\mathcal{Q}$ is a *refinement* of $\mathcal{P}$ if the point set $\mathcal{P}$ is a subset of the point set $\mathcal{Q}$.

If $\mathcal{P}, \mathcal{P}'$ are partitions of $[a, b]$ then their *common refinement* is the union of all the points of $\mathcal{P}$ and $\mathcal{P}'$.

We record now a technical lemma that plays an implicit role in several of the results that follow:

Lemma 7.1.12. *Let f be a function with domain the closed interval $[a, b]$. The Riemann integral*

$$\int_a^b f(x)\, dx$$

exists if and only if, for every $\epsilon > 0$*, there is a* $\delta > 0$ *such that if* $\mathcal{P}$ *and* $\mathcal{P}'$ *are partitions of* $[a, b]$ *with* $m(\mathcal{P}) < \delta$ *and* $m(\mathcal{P}') < \delta$ *then their common refinement* $\mathcal{Q}$ *has the property that*

$$|\mathcal{R}(f, \mathcal{P}) - \mathcal{R}(f, \mathcal{Q})| < \epsilon$$

and

$$|\mathcal{R}(f, \mathcal{P}') - \mathcal{R}(f, \mathcal{Q})| < \epsilon.$$

7.1.3 Existence of the Riemann Integral

The most important, and perhaps the simplest, fact about the Riemann integral is that a large class of familiar functions is Riemann integrable. This includes the continuous functions, the piecewise continuous functions, and more general classes of functions as well. The great classical result, which we can only touch on here, is that a function on an interval $[a, b]$ is Riemann integrable *if and only if* the set of its discontinuities has measure 0.[1] See [RUD] for all the details of this assertion.

7.1.4 Integrability of Continuous Functions

We now formalize the preceding discussion.

Theorem 7.1.13. *Let* f *be a continuous function on a non-empty closed interval* $[a, b]$*. Then* f *is Riemann integrable on* $[a, b]$*. That is to say,* $\int_a^b f(x)\,dx$ *exists.*

Example 7.1.14. We can be sure that the integral

$$\int_0^1 e^{-x^2}\,dx$$

exists (just because the integrand is continuous), even though this integral is impossible to compute by hand in closed form. □

We next note an important fact about Riemann integrable functions. A Riemann integrable function on an interval $[a, b]$ *must be bounded.* If it were not, then one could choose the points s_j in the construction of $\mathcal{R}(f, \mathcal{P})$ so that $f(s_j)$ is arbitrarily large, and the Riemann sums would become arbitrarily large, hence cannot converge.

Having said this, we do note that there is a theory of *improper integrals* that allows for integation of some unbounded functions. Technically these integrals are not Riemann integrals, but they are limits of Riemann integrals.

[1]Here a set S has measure zero if, for any $\epsilon > 0$, S can be covered by a union of open intervals the sum of whose lengths is less than ϵ.

7.2 Properties of the Riemann Integral

7.2.1 Existence Theorems

We begin this section with a few elementary properties of the integral that reflect its linear nature.

Theorem 7.2.1. *Let $[a, b]$ be a non-empty interval, let f and g be Riemann integrable functions on the interval, and let α be a real number. Then $f \pm g$ and $\alpha \cdot f$ are integrable and we have*

1. $\displaystyle\int_a^b f(x) \pm g(x)\, dx = \int_a^b f(x)\, dx \pm \int_a^b g(x)\, dx;$

2. $\displaystyle\int_a^b \alpha \cdot f(x)\, dx = \alpha \cdot \int_a^b f(x)\, dx;$

Theorem 7.2.2. *If c is a point of the interval $[a, b]$ and if f is Riemann integrable on both $[a, c]$ and $[c, b]$ then f is integrable on $[a, b]$ and*

$$\int_a^c f(x)dx + \int_c^b f(x)\, dx = \int_a^b f(x)\, dx.$$

Remark 7.2.3. If we adopt the convention that

$$\int_b^a f(x)\, dx = -\int_a^b f(x)\, dx$$

(which is consistent with the way that the integral was defined in the first place), then Theorem 7.2.2 is true even when c is not an element of $[a, b]$. For instance, suppose that $c < a < b$. Then, by Theorem 7.2.2,

$$\int_c^a f(x)\, dx + \int_a^b f(x)\, dx = \int_c^b f(x)\, dx\, .$$

But this may be rearranged to read

$$\int_a^b f(x)\, dx = -\int_c^a f(x)\, dx + \int_c^b f(x)\, dx = \int_a^c f(x)\, dx + \int_c^b f(x)\, dx\, .$$

EXAMPLE 7.2.4. Suppose that we know that

$$\int_0^4 f(x)\, dx = 3 \qquad \text{and} \qquad \int_2^4 f(x)\, dx = -5\, .$$

Then we may conclude that

$$\int_0^2 f(x)\, dx = \int_0^4 f(x)\, dx - \int_2^4 f(x)\, dx = 3 - (-5) = 8\, .$$

□

7.2.2 INEQUALITIES FOR INTEGRALS

One of the basic techniques of analysis is to perform estimates. Thus we require certain fundamental inequalities about integrals. These are recorded in the next theorem.

Theorem 7.2.5. *Let f and g be integrable functions on a non-empty interval $[a, b]$. Then*

(7.2.5.1) $$\left| \int_a^b f(x)\, dx \right| \le \int_a^b |f(x)|\, dx;$$

(7.2.5.2) *If $f(x) \le g(x)$ for all $x \in [a, b]$ then* $\int_a^b f(x)\, dx \le \int_a^b g(x)\, dx.$

EXAMPLE 7.2.6. We may estimate that

$$\int_0^{\pi/2} x \cdot \sin x\, dx \le \frac{\pi}{2} \int_0^{\pi/2} \sin x\, dx = \frac{\pi}{2}.$$

Likewise,

$$\int_1^e \frac{\ln x}{x^2}\, dx \le 1 \cdot \int_1^e \frac{\ln x}{x}\, dx = \frac{1}{2}.$$

Finally,

$$\int_1^\pi \left| \frac{\cos x}{x} \right| dx \le \int_1^\pi \frac{1}{x}\, dx = \ln \pi .$$

□

Lemma 7.2.7. *If f is a Riemann integrable function on $[a, b]$ and if ϕ is a continuous function on a compact interval that contains the range of f then $\phi \circ f$ is Riemann integrable.*

Corollary 7.2.8. *If f and g are Riemann integrable on $[a, b]$, then so is the function $f \cdot g$.*

7.2.3 PRESERVATION OF INTEGRABLE FUNCTIONS UNDER COMPOSITION

The following result is the so-called "change of variables formula". In some calculus books it is also referred to as the "u-substitution." This device is useful for transforming a given integral into another (on a different domain) that may be easier to handle.

Theorem 7.2.9. *Let f be an integrable function on an interval $[a, b]$ of positive length. Let ψ be a continuously differentiable function from another interval $[\alpha, \beta]$ of positive length into $[a, b]$. Assume that ψ is monotone increasing, one-to-one, and onto. Then*

$$\int_a^b f(x)\,dx = \int_\alpha^\beta f(\psi(t)) \cdot \psi'(t)\,dt.$$

EXAMPLE 7.2.10. Let $f(x) = (\sin x^2) \cdot 2x$ on the interval $[0, \pi]$. Let $\psi(t) = \sqrt{t}$. According to the theorem, then,

$$\begin{aligned}
\int_0^\pi \sin x^2 \cdot 2x\,dx &= \int_0^\pi f(x)\,dx \\
&= \int_0^{\pi^2} f(\psi(t)) \cdot \psi'(t)\,dt \\
&= \int_0^{\pi^2} \left[(\sin t) \cdot 2\sqrt{t}\right] \cdot \frac{1}{2\sqrt{t}}\,dt \\
&= \int_0^{\pi^2} \sin t\,dt \\
&= -\cos \pi^2 + \cos 0 \\
&= -\cos \pi^2 + 1\,.
\end{aligned}$$

EXAMPLE 7.2.11. Let $f(x) = e^{1/x}/x^2$ on the interval $[1/2, 1]$. Let $\psi(t) = 1/t$. According to the theorem,

$$\begin{aligned}
\int_{1/2}^1 \frac{e^{1/x}}{x^2}\,dx &= \int_{1/2}^1 f(x)\,dx \\
&= \int_2^1 f(\psi(t))\psi'(t)\,dt \\
&= \int_2^1 e^t \cdot t^2 \cdot \frac{-1}{t^2}\,dt \\
&= \int_1^2 e^t\,dt \\
&= e^2 - e\,.
\end{aligned}$$

7.2.4 THE FUNDAMENTAL THEOREM OF CALCULUS

Theorem 7.2.12. *Let f be an integrable function on the interval $[a, b]$. For $x \in [a, b]$ we define*

$$F(x) = \int_a^x f(s)ds.$$

If f is continuous at $x \in (a, b)$ then

$$F'(x) = f(x).$$

We conclude with this important interpretation of the fundamental theorem:

Corollary 7.2.13. *If f is a continuous function on $[a, b]$ and if G is any continuously differentiable function on $[a, b]$ whose derivative equals f on (a, b) then*

$$\int_a^b f(x)\, dx = G(b) - G(a).$$

EXAMPLE 7.2.14. Let us calculate

$$\frac{d}{dx} \int_{x^2}^{x^3} \sin(\ln t)\, dt\,. \tag{7.2.14.1}$$

It is useful to let $G(t)$ be an antiderivative of the function $\sin(\ln t)$. So

$$G(t) = \int_1^t \sin(\ln s)\, ds\,.$$

Then the expression (7.2.14.1) may be rewritten as

$$\frac{d}{dx}\Big[G(x^3) - G(x^2)\Big].$$

This is something that we can calculate using the chain rule. The result is that

$$(7.2.14.1) = G'(x^3)\cdot 3x^2 - G'(x^2)\cdot 2x = \sin(\ln(x^3))\cdot 3x^2 - \sin(\ln(x^2))\cdot 2x\,.$$

7.2.5 MEAN VALUE THEOREMS

Like the derivative, the integral enjoys several "mean value" properties. In fact the integral is more robust than the derivative, so there is more that one can say in this context. We shall state just two versions of the mean value theorem for integrals.

First Mean Value Theorem *Let f be a continuous function on $[a, b]$. Then there is a number $\xi \in [a, b]$ such that*

$$f(\xi) = \frac{1}{b-a} \int_a^b f(t)\, dt\,.$$

This is a useful fact, and almost obvious. For the expression on the right, being the average of f over the interval, must lie between the maximum value of f on $[a, b]$ and the minimum value of f on $[a, b]$. By the intermediate value property for continuous functions, ξ must therefore exist.

Second Mean Value Theorem *Let f be a continuous function on $[a, b]$ and let ψ be a positive and integrable function on $[a, b]$. Then there is a $\xi \in [a, b]$ such that*

$$\int_a^b f(t)\psi(t)\, dt = f(\xi) \int_a^b \psi(t)\, dt\,.$$

The proof of this second mean value theorem is similar to that of the first.

7.3 FURTHER RESULTS ON THE RIEMANN INTEGRAL

7.3.1 THE RIEMANN-STIELTJES INTEGRAL

Fix an interval $[a, b]$ and a monotonically increasing function α on $[a, b]$. If $\mathcal{P} = \{p_0, p_1, \ldots, p_k\}$ is a partition of $[a, b]$, let $\Delta\alpha_j = \alpha(p_j) - \alpha(p_{j-1})$. Let f be a bounded function on $[a, b]$ and define *the upper Riemann sum* of f with respect to α and the *lower Riemann sum* of f with respect to α as follows:

$$\mathcal{U}(f, \mathcal{P}, \alpha) = \sum_{j=1}^{k} M_j \Delta\alpha_j$$

and

$$\mathcal{L}(f, \mathcal{P}, \alpha) = \sum_{j=1}^{k} m_j \Delta\alpha_j.$$

The notation M_j denotes the supremum of f on the interval $I_j = [p_{j-1}, p_j]$ and m_j denotes the infimum of f on I_j.

In the special case $\alpha(x) = x$ these Riemann sums have a form similar to the Riemann sums considered in the first two sections. Moreover, in this special case,

$$\mathcal{L}(f, \mathcal{P}, \alpha) \le \mathcal{R}(f, \mathcal{P}) \le \mathcal{U}(f, \mathcal{P}, \alpha).$$

Returning to general α, we define

$$I^*(f) = \inf \mathcal{U}(f, \mathcal{P}, \alpha)$$

and

$$I_*(f) = \sup \mathcal{L}(f, \mathcal{P}, \alpha).$$

The supremum and infimum are taken with respect to all partitions of the interval $[a, b]$. These are, respectively, the *upper* and *lower integrals* of f with respect to α on $[a, b]$.

By definition it is always true that, for any partition $\mathcal{P}$,

$$\mathcal{L}(f, \mathcal{P}, \alpha) \leq I_*(f) \leq I^*(f) \leq \mathcal{U}(f, \mathcal{P}, \alpha). \tag{7.3.1.1}$$

It is natural to declare the integral to exist when the upper and lower integrals agree:

Definition 7.3.1. Let α be a monotone increasing function on the interval $[a, b]$ and let f be a bounded function on $[a, b]$. We say that the *Riemann-Stieltjes integral of f with respect to α* exists if

$$I^*(f) = I_*(f).$$

When the integral exists we denote it by

$$\int_a^b f \, d\alpha.$$

The definition of Riemann-Stieltjes integral differs from the definition of Riemann integral that we used in the preceding sections. It turns out that, when $\alpha(x) = x$, then the two definitions are equivalent. In the present generality it is easier to deal with upper and lower integrals in order to determine the existence of integrals. We now repeat an essential definition.

Definition 7.3.2. Let $\mathcal{P}$ and $\mathcal{Q}$ be partitions of the interval $[a, b]$. If each point of $\mathcal{P}$ is also an element of $\mathcal{Q}$ then we call $\mathcal{Q}$ a *refinement* of $\mathcal{P}$.

The refinement $\mathcal{Q}$ is obtained by adding points to $\mathcal{P}$. The mesh of $\mathcal{Q}$ will be less than or equal to that of $\mathcal{P}$. The following lemma enables us to deal effectively with our new language:

Lemma 7.3.3. *Let $\mathcal{P}$ be a partition of the interval $[a, b]$ and f a function on $[a, b]$. Fix a monotone increasing function α on $[a, b]$. If $\mathcal{Q}$ is a refinement of $\mathcal{P}$ then*

$$\mathcal{U}(f, \mathcal{Q}, \alpha) \leq \mathcal{U}(f, \mathcal{P}, \alpha)$$

and

$$\mathcal{L}(f, \mathcal{Q}, \alpha) \geq \mathcal{L}(f, \mathcal{P}, \alpha).$$

EXAMPLE 7.3.4. Let $[a, b] = [0, 10]$ and let $\alpha(x)$ be the *greatest integer function*. That is, $\alpha(x)$ is the greatest integer that does not exceed x. So, for example, $\alpha(0.5) = 0, \alpha(2) = 2$, and $\alpha(-3/2) = -2$. Certainly α is a monotone increasing function on $[0, 10]$. Let f be any continuous function on $[0, 10]$. We shall determine whether

$$\int_0^{10} f \, d\alpha$$

exists and, if it does, calculate its value.

Let $\mathcal{P} = \{p_0, p_1, \ldots, p_k\}$ be a partition of $[0, 10]$. By the lemma, it is to our advantage to assume that the mesh of $\mathcal{P}$ is smaller than 1. Observe that $\Delta\alpha_j$ equals the number of integers that lie in the interval I_j—that is, either 0 or 1. Let $I_{j_0}, I_{j_2}, \ldots I_{j_{10}}$ be the intervals from the partition that do in fact contain integers (the first of these contains 0, the second contains 1, and so on up to 10). Then

$$\mathcal{U}(f, \mathcal{P}, \alpha) = \sum_{\ell=0}^{10} M_{j_\ell} \Delta\alpha_{j_\ell} = \sum_{\ell=1}^{10} M_{j_\ell}$$

and

$$\mathcal{L}(f, \mathcal{P}, \alpha) = \sum_{\ell=0}^{10} m_{j_\ell} \Delta\alpha_{j_\ell} = \sum_{\ell=1}^{10} m_{j_\ell}$$

because the increment $\Delta\alpha_{j_\ell}$ for an interval containing an integer will be 1 (and for an interval not containing an integer the increment of course will be 0). Notice, for instance, that $\Delta\alpha_{j_0} = 0$ since $\alpha(0) = \alpha(p_1) = 0$. But $\Delta\alpha_{j_1} = 1$.

Let $\epsilon > 0$. Since f is uniformly continuous on $[0, 10]$, we may choose a $\delta > 0$ such that $|s - t| < \delta$ implies that $|f(s) - f(t)| < \epsilon/20$. If $m(\mathcal{P}) < \delta$ then it follows that $|f(\ell) - M_{j_\ell}| < \epsilon/20$ and $|f(\ell) - m_{j_\ell}| < \epsilon/20$ for $\ell = 0, 1, \ldots, 10$. Therefore

$$\mathcal{U}(f, \mathcal{P}, \alpha) < \sum_{\ell=1}^{10} \left(f(\ell) + \frac{\epsilon}{20} \right)$$

and

$$\mathcal{L}(f, \mathcal{P}, \alpha) > \sum_{\ell=1}^{10} \left(f(\ell) - \frac{\epsilon}{20} \right).$$

Rearranging the first of these inequalities leads to

$$\mathcal{U}(f, \mathcal{P}, \alpha) < \left(\sum_{\ell=1}^{10} f(\ell) \right) + \frac{\epsilon}{2}$$

and likewise we may obtain that

$$\mathcal{L}(f, \mathcal{P}, \alpha) > \left(\sum_{\ell=1}^{10} f(\ell) \right) - \frac{\epsilon}{2}.$$

Thus, since I_* and I^* are trapped between $\mathcal{U}$ and $\mathcal{L}$, we conclude that

$$|I_*(f) - I^*(f)| < \epsilon.$$

We have seen that, if the partition is fine enough, then the upper and lower integrals of f with respect to α differ by at most ϵ. It follows that $\int_0^{10} f \, d\alpha$ exists. Moreover,

$$\left| I^*(f) - \sum_{\ell=1}^{10} f(\ell) \right| < \epsilon$$

and

$$\left| I_*(f) - \sum_{\ell=1}^{10} f(\ell) \right| < \epsilon.$$

We conclude that

$$\int_0^{10} f \, d\alpha = \sum_{\ell=1}^{10} f(\ell).$$

□

The example demonstrates that the language of the Riemann-Stieltjes integral allows us to think of the integral in a concrete fashion as a generalization of the summation process. This is frequently useful, both philosophically and for practical reasons.

7.3.2 RIEMANN'S LEMMA

The next result, sometimes called Riemann's Lemma, is crucial for proving the existence of Riemann-Stieltjes integrals.

Proposition 7.3.5. *Let α be a monotone increasing function on $[a, b]$ and f a bounded function on the interval. The Riemann-Stieltjes integral of f with respect to α exists if and only if, for every $\epsilon > 0$, there is a partition $\mathcal{P}$ such that*

$$|\mathcal{U}(f, \mathcal{P}, \alpha) - \mathcal{L}(f, \mathcal{P}, \alpha)| < \epsilon. \tag{7.3.5.1}$$

We note in passing that the basic properties of the Riemann integral noted in Section 7.2 (Theorems 7.2.1 and 7.2.2) hold without change for the Riemann-Stieltjes integral.

7.4 Advanced Results on Integration Theory

7.4.1 Existence for the Riemann-Stieltjes Integral

We now turn to enunciating the existence of certain Riemann-Stieltjes integrals.

Theorem 7.4.1. *Let f be continuous on $[a, b]$ and assume that α is monotonically increasing. Then*

$$\int_a^b f \, d\alpha$$

exists.

Theorem 7.4.2. *If α is a monotone increasing and continuous function on the interval $[a, b]$ and if f is monotonic on $[a, b]$ then $\int_a^b f \, d\alpha$ exists.*

7.4.2 Integration by Parts

One of the useful features of Riemann-Stieltjes integration is that it puts integration by parts into a very natural setting. We begin with a lemma:

Lemma 7.4.3. *Let f be continuous on an interval $[a, b]$ and let g be monotone increasing and continuous on that interval. If G is an antiderivative for g then*

$$\int_a^b f(x)g(x) \, dx = \int_a^b f \, dG.$$

Theorem 7.4.4. *Suppose that both f and g are continuous, monotone increasing functions on the interval $[a, b]$. Let F be an antiderivative for f on $[a, b]$ and G an antiderivative for g on $[a, b]$. Then we have*

$$\int_a^b F \, dG = [F(b) \cdot G(b) - F(a) \cdot G(a)] - \int_a^b G \, dF$$

Example 7.4.5. We may apply integration by parts to the integral

$$I = \int_0^\pi x \cdot \cos x \, dx \, .$$

The result is

$$I = [x \sin x]_0^\pi - \int_0^\pi \sin x \, dx = -2 \, .$$

□

Remark 7.4.6. The integration by parts formula can be proved by applying *summation by parts* (Subsection 3.3.2) to the Riemann sums for the integral

$$\int_a^b F\,dG\,.$$

7.4.3 LINEARITY PROPERTIES

We have already observed that the Riemann-Stieltjes integral

$$\int_a^b f\,d\alpha$$

is linear in f; that is,

$$\int_a^b (f+g)\,d\alpha = \int_a^b f\,d\alpha + \int_a^b g\,d\alpha$$

and

$$\int_a^b c\cdot f\,d\alpha = c\cdot\int_a^b f\,d\alpha$$

when both f and g are Riemann-Stieltjes integrable with respect to α and for any constant c. We also would expect, from the very way that the integral is constructed, that it would be linear in the α entry. But we have not even defined the Riemann-Stieltjes integral for non-increasing α. And what of a function α that is the difference of two monotone increasing functions? Such a function certainly need not be monotone. Is it possible to identify which functions α can be decomposed as sums or differences of monotonic functions? It turns out that there is a satisfactory answer to these questions, and we now discuss these matters briefly.

7.4.4 BOUNDED VARIATION

Definition 7.4.7. If α is a monotonically *decreasing* function on $[a,b]$ and f is a function on $[a,b]$ then we define

$$\int_a^b f\,d\alpha = -\int_a^b f d(-\alpha)$$

when the right side exists.

The definition exploits the simple observation that if α is monotone decreasing then $-\alpha$ is monotone increasing; hence the preceding theory applies to the function $-\alpha$.

Next we have

Definition 7.4.8. Let α be a function on $[a, b]$ that can be expressed as

$$\alpha(x) = \alpha_1(x) - \alpha_2(x),$$

where both α_1 and α_2 are monotone increasing. Then for any f on $[a, b]$ we define

$$\int_a^b f\, d\alpha = \int_a^b f\, d\alpha_1 - \int_a^b f\, d\alpha_2,$$

provided that both integrals on the right exist.

Now, by the very way that we have formulated our definitions, $\int_a^b f\, d\alpha$ is linear in both the f entry and the α entry. But the definitions are not satisfactory unless we can identify those α that can actually occur in the last definition. This leads us to a new class of functions.

Definition 7.4.9. Let f be a function on the interval $[a, b]$. For $x \in [a, b]$ we define

$$Vf(x) = \sup \sum_{j=1}^{k} \left| f(p_j) - f(p_{j-1}) \right|,$$

where the supremum is taken over all partitions $\mathcal{P} = \{p_0, p_1, \ldots, p_k\}$ of the interval $[a, x]$.

If $Vf \equiv Vf(b) < \infty$ then the function f is said to be of *bounded variation* on the interval $[a, b]$. In this circumstance the quantity $Vf(b)$ is called the *total variation* of f on $[a, b]$.

A function of bounded variation has the property that its graph does not have unbounded total oscillation.

EXAMPLE 7.4.10. Define $f(x) = \sin x$, with domain the interval $[0, 2\pi]$. Let us calculate Vf. Let $\mathcal{P}$ be a partition of $[0, 2\pi]$. Since adding points to the partition only makes the sum

$$\sum_{j=1}^{k} \left| f(p_j) - f(p_{j-1}) \right|$$

larger (by the triangle inequality), we may as well suppose that $\mathcal{P} = \{p_0, p_1, p_2, \ldots, p_k\}$ contains the points $\pi/2$, $3\pi/2$. Say that $p_{\ell_1} = \pi/2$

and $p_{\ell_2} = 3\pi/2$. Then

$$\begin{aligned}\sum_{j=1}^{k} |f(p_j) - f(p_{j-1})| &= \sum_{j=1}^{\ell_1} |f(p_j) - f(p_{j-1})| \\ &\quad + \sum_{j=\ell_1+1}^{\ell_2} |f(p_j) - f(p_{j-1})| \\ &\quad + \sum_{j=\ell_2+1}^{k} |f(p_j) - f(p_{j-1})| .\end{aligned}$$

However, f is monotone increasing on the interval $[0, \pi/2] = [0, p_{\ell_1}]$. Therefore the first sum is just

$$\sum_{j=1}^{\ell_1} f(p_j) - f(p_{j-1}) = f(p_{\ell_1}) - f(p_0) = f(\pi/2) - f(0) = 1.$$

Similarly, f is monotone on the intervals $[\pi/2, 3\pi/2] = [p_{\ell_1}, p_{\ell_2}]$ and $[3\pi/2, 2\pi] = [p_{\ell_2}, p_k]$. Thus the second and third sums equal $f(p_{\ell_1}) - f(p_{\ell_2}) = 2$ and $f(p_k) - f(p_{\ell_2}) = 1$ respectively. It follows that

$$Vf = Vf(2\pi) = 1 + 2 + 1 = 4.$$

Of course $Vf(x)$ for any $x \in [0, 2\pi]$ can be computed by similar means.

In general, if f is a continuously differentiable function on an interval $[a, b]$ then

$$Vf(x) = \int_a^x |f'(t)| dt.$$

□

Lemma 7.4.11. *Let f be a function of bounded variation on the interval $[a, b]$. Then the function Vf is monotone increasing on $[a, b]$.*

Lemma 7.4.12. *Let f be a function of bounded variation on the interval $[a, b]$. Then the function $Vf - f$ is monotone increasing on the interval $[a, b]$.*

Now we may combine the last two lemmas to obtain our main result:

Proposition 7.4.13. *If a function f is of bounded variation on $[a, b]$, then f may be written as the difference of two monotone increasing functions. Namely,*

$$f = Vf - [Vf - f].$$

Conversely, the difference of two monotone increasing functions is a function of bounded variation.

The main point of this discussion is the following theorem:

Theorem 7.4.14. *If f is a continuous function on $[a, b]$ and if α is of bounded variation on $[a, b]$ then the integral*

$$\int_a^b f \, d\alpha$$

exists.

If g is of bounded variation on $[a, b]$ and if β is a continuous function of bounded variation on $[a, b]$ then the integral

$$\int_a^b g \, d\beta$$

exists.

Both of these results follow by expressing the function of bounded variation as the difference of two monotone functions—according to Proposition 7.4.13.

CHAPTER 8

Sequences and Series of Functions

8.1 Partial Sums and Pointwise Convergence

8.1.1 Sequences of Functions

A *sequence of functions* is usually written

$$f_1(x), f_2(x), \ldots \quad \text{or} \quad \{f_j(x)\}_{j=1}^{\infty} \quad \text{or} \quad \{f_j\} .$$

We will generally assume that the functions f_j all have the same domain S.

Definition 8.1.1. A sequence of functions $\{f_j\}_{j=1}^{\infty}$ with domain $S \subseteq \mathbb{R}$ is said to *converge pointwise* to a limit function f on S if, for each $x \in S$, the sequence of numbers $\{f_j(x)\}$ converges to $f(x)$. We write $\lim_{j\to\infty} f_j(x) = f(x)$.

EXAMPLE 8.1.2. Define $f_j(x) = x^j$ with domain $S = \{x : 0 \le x \le 1\}$. If $0 \le x < 1$ then $f_j(x) \to 0$. However, $f_j(1) \to 1$. Therefore the sequence f_j converges to the function

$$f(x) = \begin{cases} 0 & \text{if} \quad 0 \le x < 1 \\ 1 & \text{if} \quad x = 1 \end{cases}$$

See Figure 8.1. □

Here are some of the basic questions that we must ask about a sequence of functions f_j that converges to a function f on a domain S :

1. If the functions f_j are continuous then is f continuous?

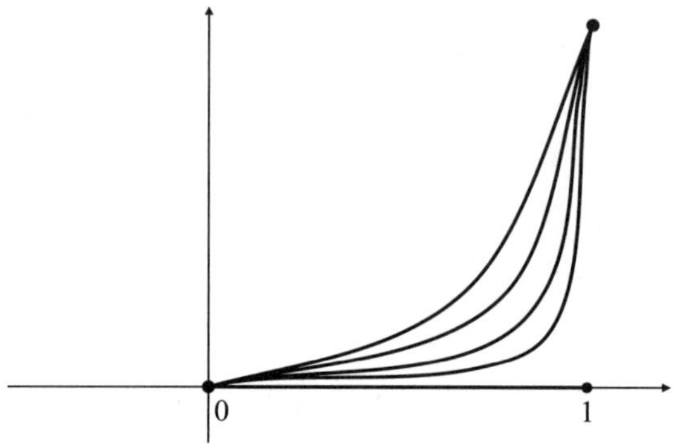

FIGURE 8.1. Pointwise convergence of the sequence $\{x^j\}$.

2. If the functions f_j are integrable on an interval I then is f integrable on I? If f is integrable on I then does the sequence $\int_I f_j(x)\,dx$ converge to $\int_I f(x)dx$?

3. If the functions f_j are differentiable then is f differentiable? If f is differentiable then does the sequence f_j' converge to f'?

8.1.2 UNIFORM CONVERGENCE

We see from Example 8.1.2 that the answer to the first question of the last subsection is "no ": Each of the f_j is continuous but f certainly is not. It turns out that, in order to obtain a favorable answer to our questions, we must consider a stricter notion of convergence of functions. This motivates the next definition.

Definition 8.1.3. Let f_j be a sequence of functions on a domain S. We say that the functions f_j converge *uniformly* to f on S if, given $\epsilon > 0$, there is an $N > 0$ such that, for any $j > N$ and any $s \in S$, it holds that $|f_j(s) - f(s)| < \epsilon$.

The special feature of uniform convergence is that the rate at which $f_j(s)$ converges is independent of $s \in S$. In Example 8.1.2, $f_j(x)$ is converging very rapidly to zero for x near zero but arbitrarily slowly to zero for x near 1 (draw a sketch to help you understand this point). In the next example we shall establish this assertion rigorously:

EXAMPLE 8.1.4. The sequence $f_j(x) = x^j$ does not converge uniformly to the limit function

$$f(x) = \begin{cases} 0 \text{ if } 0 \le x < 1 \\ 1 \text{ if } x = 1 \end{cases}$$

on the domain $S = [0, 1]$. In fact it does not even do so on the smaller domain $[0, 1)$. Again see Figure 8.1.

To see this, notice that no matter how large j is we have, by the Mean Value Theorem, that

$$f_j(1) - f_j(1 - 1/(2j)) = \frac{1}{2j} \cdot f_j'(\xi)$$

for some ξ between $1-1/(2j)$ and 1. But $f_j'(x) = j \cdot x^{j-1}$ hence $|f_j'(\xi)| < j$ and we conclude that

$$|f_j(1) - f_j\big(1 - 1/(2j)\big)| < \frac{1}{2}$$

or

$$f_j\big(1 - 1/(2j)\big) > f_j(1) - \frac{1}{2} = \frac{1}{2}.$$

In conclusion, no matter how large j, there will be values of x (namely $x = 1 - 1/(2j)$) at which $f_j(x)$ is at least distance $1/2$ from the limit 0. We conclude that the convergence is not uniform. □

Theorem 8.1.5. *If f_j are continuous functions on a set S that converge uniformly on S to a function f then f is also continuous.*

Next we turn our attention to integration.

EXAMPLE 8.1.6. Define functions

$$f_j(x) = \begin{cases} 0 \text{ if } x = 0 \\ j \text{ if } 0 < x \leq 1/j \\ 0 \text{ if } 1/j < x \leq 1 . \end{cases}$$

Then $\lim_{j\to\infty} f_j(x) = 0$ for all x in the interval $I = [0, 1]$. However

$$\int_0^1 f_j(x)\,dx = \int_0^{1/j} j\,dx = 1$$

for every j. Thus the f_j converge to the integrable limit function $f(x) \equiv 0$, but their integrals do not converge to the integral of f. □

EXAMPLE 8.1.7. Let $q_1, q_2, \ldots$ be an enumeration of the rationals in the interval $I = [0, 1]$. Define functions

$$f_j(x) = \begin{cases} 1 \text{ if } x \in \{q_1, q_2, \ldots, q_j\} \\ 0 \text{ if } x \notin \{q_1, q_2, \ldots, q_j\} \end{cases}$$

Then the functions f_j converge pointwise to the Dirichlet function f that is equal to 1 on the rationals and 0 on the irrationals. Each of the functions f_j has integral 0 on I. But the function f is not integrable on I. □

The last two examples show that something more than pointwise convergence is needed in order for the integral to respect the limit process.

Theorem 8.1.8. *Let f_j be integrable functions on a bounded interval $[a, b]$ and suppose that the functions f_j converge uniformly to the limit function f. Then f is integrable on $[a, b]$ and*

$$\lim_{j\to\infty} \int_a^b f_j(x)\,dx = \int_a^b f(x)\,dx\,.$$

We have succeeded in answering questions **1** and **2** that were raised at the beginning of the section. In the next section we shall answer question **3**.

8.2 More on Uniform Convergence

8.2.1 Commutation of Limits

In general limits do not commute. Since the integral is defined with a limit, and since we saw in the last section that integrals do not always respect limits of functions, we know some concrete instances of non-commutation of limits. The fact that continuity is defined with a limit, and that the limit of continuous functions need not be continuous, gives even more examples of situations in which limits do not commute. Let us now turn to a situation in which limits *do* commute:

Theorem 8.2.1. *Fix a set S and a point $s \in S$. Assume that the functions f_j converge uniformly on the domain $S \setminus \{s\}$ to a limit function f. Suppose that each function $f_j(x)$ has a limit as $x \to s$. Then f itself has a limit as $x \to s$ and*

$$\lim_{x\to s} f(x) = \lim_{j\to\infty} \lim_{x\to s} f_j(x).$$

Because of the way that f is defined, we may rewrite this conclusion as

$$\lim_{x\to s} \lim_{j\to\infty} f_j(x) = \lim_{j\to\infty} \lim_{x\to s} f_j(x).$$

In other words, the limits in x and in j commute.

EXAMPLE 8.2.2. Consider the limit

$$\lim_{x\to 1^-} \lim_{j\to\infty} x^j\,.$$

This is easily seen to equal 0. But

$$\lim_{j\to\infty} \lim_{x\to 1^-} x^j$$

equals 1. The reason that these two limits are unequal is that the convergence of x^j is not uniform (See Example 8.1.4).

By contrast, the limit

$$\lim_{x\to\pi}\lim_{j\to\infty}\frac{\sin jx}{j}$$

can be calculated in any order (because the functions converge uniformly). The limit is equal to zero. □

8.2.2 THE UNIFORM CAUCHY CONDITION

Parallel with our notion of Cauchy sequence of numbers, we have a concept of Cauchy sequence of functions in the uniform sense:

Definition 8.2.3. A sequence of functions f_j on a domain S is called *a uniformly Cauchy sequence* if, for each $\epsilon > 0$, there is an $N > 0$ such that if $j, k > N$ then

$$\left|f_j(s) - f_k(s)\right| < \epsilon \quad \forall s \in S.$$

Proposition 8.2.4. *A sequence of function f_j is uniformly Cauchy on a domain S if and only if the sequence converges uniformly to a limit function f on the domain S.*

We will use the last two results in our study of the limits of differentiable functions. First we consider an example.

EXAMPLE 8.2.5. Define the function

$$f_j(x) = \begin{cases} 0 & \text{if } x \le 0 \\ jx^2 & \text{if } 0 < x \le 1/(2j) \\ x - 1/(4j) & \text{if } 1/(2j) < x < \infty \end{cases}$$

We leave it as an exercise to check that the functions f_j converge uniformly on the entire real line to the function

$$f(x) = \begin{cases} 0 \text{ if } x \le 0 \\ x \text{ if } x > 0 \end{cases}$$

(draw a sketch to help you see this). Each of the functions f_j is continuously differentiable on the entire real line, but f is not differentiable at 0. □

8.2.3 LIMITS OF DERIVATIVES

It turns out that we must strengthen our convergence hypotheses if we want the limit process to respect differentiation. The basic result is

Theorem 8.2.6. *Suppose that a sequence f_j of differentiable functions on an open interval I converges pointwise to a limit function f. Suppose further that the differentiated sequence f_j' converges uniformly on I to a limit function g. Then the limit function f is differentiable on I and $f'(x) = g(x)$ for all $x \in I$.*

Remark 8.2.7. A little additional effort shows that we need only assume in the theorem that the functions f_j converge at a single point x_0 in the domain. Then the conclusion is that the limit function is differentiable at x_0 and has the predictable derivative at that point.

EXAMPLE 8.2.8. Consider the sequence

$$f_j(x) = \frac{\sin j^2 x}{j}.$$

These functions converge uniformly to 0 on the entire real line. But their derivatives do not converge. Check for yourself to see that the key hypothesis of Theorem 8.2.6 fails for this example. Draw a sketch of f_2 and f_4 so that you can see what is going on. The conclusion of the theorem fails dramatically at every point. □

8.3 SERIES OF FUNCTIONS

8.3.1 SERIES AND PARTIAL SUMS

Definition 8.3.1. The formal expression

$$\sum_{j=1}^{\infty} f_j(x),$$

where the f_j are functions on a common domain S, is called a *series of functions.* For $N = 1, 2, 3, \ldots$ the expression

$$S_N(x) = \sum_{j=1}^{N} f_j(x) = f_1(x) + f_2(x) + \cdots + f_N(x)$$

is called the Nth *partial sum* for the series. In case

$$\lim_{N\to\infty} S_N(x)$$

exists and is finite we say that the series *converges* at x. Otherwise we say that the series *diverges* at x.

The question of convergence of a series of functions, which should be thought of as an *addition process*, reduces to a question about the *sequence of partial sums*. Sometimes, as in the next example, it is convenient to begin the series at some index other than $j = 1$.

EXAMPLE 8.3.2. Consider the series

$$\sum_{j=0}^{\infty} x^j .$$

This is the geometric series from Subsection 3.2.3. It converges absolutely for $|x| < 1$ and diverges otherwise.

By the formula for the partial sums of a geometric series,

$$S_N(x) = \frac{1 - x^{N+1}}{1 - x}.$$

For $|x| < 1$ we see that

$$S_N(x) \to \frac{1}{1 - x}.$$

□

8.3.2 UNIFORM CONVERGENCE OF A SERIES

Definition 8.3.3. Let

$$\sum_{j=1}^{\infty} f_j(x)$$

be a series of functions on a domain S. If the partial sums $S_N(x)$ converge uniformly on S to a limit function $g(x)$ then we say that the series *converges uniformly* on S.

Of course all of our results about uniform convergence of a *sequence* of functions translate, via the sequence of partial sums of a series, to results about uniform convergence of a *series* of functions. For example,

(a) If f_j are continuous functions on a domain S and if the series

$$\sum_{j=1}^{\infty} f_j(x)$$

converges uniformly on S to a limit function f, then f is also continuous on S.

(b) If f_j are integrable functions on $[a, b]$ and if

$$\sum_{j=1}^{\infty} f_j(x)$$

converges uniformly on $[a, b]$ to a limit function f, then f is also integrable on $[a, b]$ and

$$\int_a^b f(x)\, dx = \sum_{j=1}^{\infty} \int_a^b f_j(x)\, dx.$$

EXAMPLE 8.3.4. The series

$$\sum_{j=1}^{\infty} \frac{je^{x/j}}{2^j}$$

converges uniformly on any bounded interval $[a, b]$. The Weierstrass M-test, discussed in the next subsection, provides a means for confirming this assertion. □

Now we turn to an elegant test for uniform convergence that is due to Weierstrass.

8.3.3 THE WEIERSTRASS M-TEST

Theorem 8.3.5 (Weierstrass). *Let $\{f_j\}_{j=1}^{\infty}$ be functions on a common domain S. Assume that each $|f_j|$ is bounded on S by a constant M_j and that*

$$\sum_{j=1}^{\infty} M_j < \infty.$$

Then the series

$$\sum_{j=1}^{\infty} f_j \tag{8.3.5.1}$$

converges uniformly on the set S.

EXAMPLE 8.3.6. Let us consider the series

$$\sum_{j=1}^{\infty} 2^{-j} \sin\left(2^j x\right).$$

The sine terms oscillate so wildly that it would be difficult to calculate partial sums for this series. However, noting that the jth summand $f_j(x) = 2^{-j}\sin(2^j x)$ is dominated in absolute value by 2^{-j}, we see that the Weierstrass M–Test applies to this series. We conclude that the series converges uniformly on the entire real line.[1]

By property **(a)** of uniformly convergent series of continuous functions, we may conclude that the function f defined by our series is continuous. It is also 2π–periodic: $f(x+2\pi) = f(x)$ for every x since this assertion is true for each summand. Since the continuous function f restricted to the compact interval $[0, 2\pi]$ is uniformly continuous (Section 5.3), we may conclude that f is uniformly continuous on the entire real line.

However, it turns out that f is nowhere differentiable. The proof of this assertion follows lines similar to the treatment of nowhere differentiable functions in Subsection 6.1.3. □

Exercise: Verify the assertions of Example 8.3.6. □

8.4 THE WEIERSTRASS APPROXIMATION THEOREM

The name Weierstrass has occurred frequently in this chapter. In fact Karl Weierstrass (1815–1897) revolutionized analysis with his examples and theorems. This section is devoted to one of his most striking results. We introduce it with a motivating discussion.

It is natural to wonder whether the standard functions of calculus—$\sin x$, $\cos x$, and e^x, for instance—are actually polynomials of some very high degree. Since polynomials are so much easier to understand than these transcendental functions, an affirmative answer to this question would certainly simplify mathematics. Of course a moment's thought shows that this wish is impossible: a polynomial of degree k has at most k real roots. Since sine and cosine have infinitely many real roots they cannot be polynomials. A polynomial of degree k has the property that if it is differentiated enough times (namely $k+1$ times) then its derivative is zero. Since this is not the case for e^x, we conclude that e^x cannot be a polynomial. A similar observation holds for $\log x$.

However, in calculus we learned of a formal procedure, called Taylor series, for associating polynomials with a given function f. In some in-

[1] In fact the series converges to the Weierstrass nowhere differentiable function in Weierstrass's original formulation.

stances these polynomials form a sequence that converges back to the original function.[2] This might cause us to speculate that any reasonable function can be approximated in some fashion by polynomials. In fact the theorem of Weierstrass gives a spectacular affirmation of this supposition:

8.4.1 WEIERSTRASS'S MAIN RESULT

Theorem 8.4.1. *Let f be a continuous function on an interval $[a, b]$. Then there is a sequence of polynomials $p_j(x)$ with the property that the sequence $\{p_j\}$ converges uniformly on $[a, b]$ to f. See Figure 8.2.*

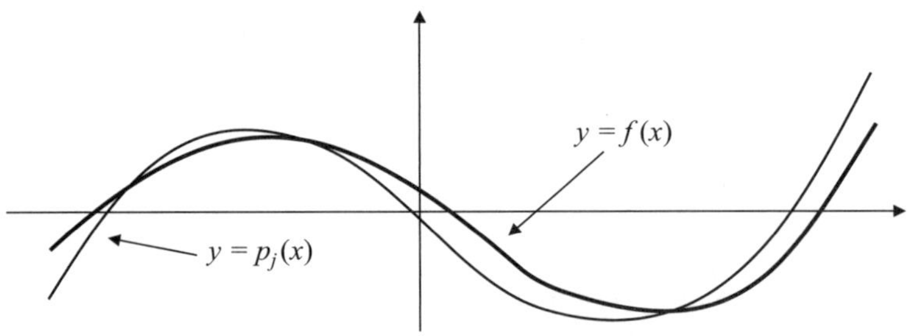

FIGURE 8.2. The Weierstrass approximation theorem.

Let us consider some consequences of the theorem. A restatement of the theorem would be that, given a continuous function f on $[a, b]$ and an $\epsilon > 0$, there is a polynomial p such that

$$|f(x) - p(x)| < \epsilon$$

for every $x \in [a, b]$. If one were programming a computer to calculate values of a fairly wild function f, the theorem guarantees that, up to a given degree of accuracy, one could use a polynomial instead (which would in fact be much easier for the computer to handle). Advanced techniques can even tell what degree of polynomial is needed to achieve a given degree of accuracy.

And notice this: Let f be the Weierstrass nowhere differentiable function. The theorem guarantees that, on any compact interval, f is the uniform limit of polynomials. Thus even the uniform limit of infinitely differentiable functions need not be differentiable—even at one point. This explains why the hypotheses of Theorem 8.2.6 needed to be so stringent.

[2]However, it must be noted that most infinitely differentiable functions f *do not* have convergent Taylor series. And, even when the series converges, it typically does not converge back to the original function f.

Remark 8.4.2. If f is a given continuous function then it is a matter of great interest to actually *produce* the polynomial that will approximate f to a pre-specified degree of accuracy. There is a large theory built around this question. Certainly the Lagrange interpolation polynomials (see [BUB] or [ABR]) will do the trick. An examination of the proof of the Weierstrass theorem that is presented in either [RUD] or [KRA1] will give another method of approximation.

EXAMPLE 8.4.3. Let f be a continuously differentiable function on the interval $[0, 1]$. Can we approximate it by polynomials p_j so that $p_j \to f$ uniformly and also $p_j' \to f'$ uniformly?

The answer is "yes." Apply Weierstrass's theorem to find polynomials q_j that converge uniformly to f'. Then integrate the q_j to produce the desired polynomials p_j. We leave the details to the reader.

If it is known that $f(1/2) = 0$ then we can produce polynomials p_j that perform the approximation described in the last two paragraphs and such that each $p_j(1/2) = 0$—just subtract a suitable constant from each polynomial. Again, details are left for the interested reader. □

CHAPTER 9

ADVANCED TOPICS

Part of the power of modern analysis is to look at things from an abstract point of view. This provides both unity and clarity. It also treats all dimensions at once. We shall endeavor to make these points clear as we proceed.

9.1 METRIC SPACES

9.1.1 THE CONCEPT OF A METRIC

This section formalizes a general context in which we may do analysis any time we have a reasonable notion of calculating distance. Such a structure will be called a metric:

Definition 9.1.1. A *metric space* is a pair (X, ρ), where X is a set and

$$\rho : X \times X \to \{t \in \mathbb{R} : t \geq 0\}$$

is a function satisfying

1. $\forall x, y \in X,\ \rho(x, y) = \rho(y, x)$;

2. $\rho(x, y) = 0$ if and only if $x = y$;

3. $\forall x, y, z \in X, \rho(x, y) \leq \rho(x, z) + \rho(z, y)$.

The function ρ is called a *metric* on X. Condition **3** is called the *triangle inequality*.

9.1.2 EXAMPLES OF METRIC SPACES

EXAMPLE 9.1.2. The pair $(\mathbb{R}, \rho)$, where $\rho(x, y) = |x - y|$, is a metric space. Each of the properties required of a metric is in this case a restatement of familiar facts from the analysis of one dimension.

The pair $(\mathbb{R}^3, \rho)$, where

$$\rho(x, y) = \|x - y\| = \sqrt{(x_1 - y_1)^2 + (x_2 - y_2)^2 + (x_3 - y_3)^2}\,,$$

is a metric space. Each of the properties required of a metric is in this case a restatement of familiar facts from the analysis of three dimensions. □

The first examples presented familiar metrics on two familiar spaces. Now we look at some new ones.

EXAMPLE 9.1.3. The pair $(\mathbb{R}^2, \rho)$, where $\rho(x, y) = \max\{|x_1 - y_1|, |x_2 - y_2|\}$, is a metric space. Only the triangle inequality is not trivial to verify; but that reduces, by consideration of several cases, to the triangle inequality of one variable.

The pair $(\mathbb{R}, \mu)$, where $\mu(x, y) = 1$ if $x \neq y$ and $= 0$ otherwise, is a metric space. Checking the triangle inequality reduces to seeing that, if $x \neq y$, then either $x \neq z$ or $y \neq z$. □

EXAMPLE 9.1.4. Let X denote the space of continuous functions on the interval $[0, 1]$. If $f, g \in X$ then let

$$\rho(f, g) = \sup_{t \in [0,1]} |f(t) - g(t)|\,.$$

Then the pair (X, ρ) is a metric space. The first two properties of a metric are obvious and the triangle inequality reduces to the triangle inequality for real numbers.

This example is a dramatic new departure from the analysis we have done in the previous eight chapters. For X is a very large space—infinite dimensional in a certain sense. Using the ideas that we are about to develop, it is nonetheless possible to study convergence, continuity, compactness, and the other basic concepts of analysis in this more general context. We shall see applications of these new techniques in later sections. □

9.1.3 CONVERGENCE IN A METRIC SPACE

Now we begin to develop the tools of analysis in metric spaces.

Definition 9.1.5. Let (X, ρ) be a metric space. A sequence $\{x_j\}$ of elements of X is said to *converge* to a point $\alpha \in X$ if, for each $\epsilon > 0$, there is an $N > 0$ such that, if $j > N$, then $\rho(x_j, \alpha) < \epsilon$. We call α the *limit* of the sequence $\{x_j\}$. We sometimes write $x_j \to \alpha$.

Compare this definition of convergence with the corresponding definition for convergence in the real line in Section 2.1. Notice that it is identical, except that the sense in which distance is measured is now more general.

EXAMPLE 9.1.6. Let (X, ρ) be the metric space from Example 9.1.4, consisting of the continuous functions on the unit interval with the indicated metric function ρ. Then $f = \sin x$ is an element of this space, and so are the functions

$$f_j = \sum_{\ell=0}^{j} (-1)^\ell \frac{x^{2\ell+1}}{(2\ell+1)!}.$$

Observe that the functions f_j are the partial sums for the Taylor series of $\sin x$. We can check from simple estimates on the error term of Taylor's theorem that the functions f_j converge uniformly to f. Thus, in the language of metric spaces, $f_j \to f$ in the metric space sense. □

9.1.4 THE CAUCHY CRITERION

Definition 9.1.7. Let (X, ρ) be a metric space. A sequence $\{x_j\}$ of elements of X is said to be *Cauchy* if, for each $\epsilon > 0$, there is an $N > 0$ such that, if $j, k > N$, then $\rho(x_j, x_k) < \epsilon$.

Now the Cauchy criterion and convergence are connected in the expected fashion:

Proposition 9.1.8. *Let $\{x_j\}$ be a convergent sequence, with limit α, in the metric space (X, ρ). Then the sequence $\{x_j\}$ is Cauchy.*

EXAMPLE 9.1.9. The converse of the proposition is true in the real numbers (with the usual metric), as we proved in Section 2.1. However, it is not true in every metric space. For example, the rationals $\mathbb{Q}$ with the usual metric $\rho(s, t) = |s - t|$ is a metric space; but the sequence

$$3, 3.1, 3.14, 3.141, 3.1415, 3.14159, \ldots,$$

while certainly Cauchy, *does not converge to a rational number.* □

Thus we are led to a definition:

9.1.5 COMPLETENESS

Definition 9.1.10. We say that a metric space (X, ρ) is *complete* if every Cauchy sequence converges to an element of the metric space.

Thus the real numbers, with the usual metric, form a complete metric space. The rational numbers do not.

EXAMPLE 9.1.11. Consider the metric space (X, ρ) from Example 9.1.4 above, consisting of the continuous functions on the closed unit interval with the indicated uniform metric function ρ. If $\{g_j\}$ is a Cauchy sequence in this metric space then each g_j is a continuous function on the unit interval and this sequence of continuous functions is Cauchy in the uniform sense (see Chapter 6). Therefore they converge uniformly to a limit function g that must be continuous. We conclude that the metric space (X, ρ) is complete. □

EXAMPLE 9.1.12. Consider the metric space (X, ρ) consisting of the polynomials, taken to have domain the interval $[0, 1]$, with the distance function $\rho(f, g) = \sup_{t\in[0,1]} |f(t) - g(t)|$. This metric space is *not* complete. For if h is any continuous function on $[0, 1]$ that is not a polynomial, such as $h(x) = \sin x$, then by the Weierstrass Approximation Theorem there is a sequence $\{p_j\}$ of polynomials that converges uniformly on $[0, 1]$ to h. Thus this sequence $\{p_j\}$ will be Cauchy in the metric space, but it *does not converge to an element of the metric space.* We conclude that the metric space (X, ρ) is not complete. □

9.1.6 ISOLATED POINTS

If (X, ρ) is a metric space, $P \in X$, and $r > 0$ then we let $B(P, r) \equiv \{x \in X : \rho(x, P) < r\}$. It is also sometimes useful to let $\overline{B}(P, r) = \{x \in X : \rho(x, P) \leq r\}$. These sets are called, respectively, the *open ball* with center P and radius r and the *closed ball* with center P and radius r.

Definition 9.1.13. Let (X, ρ) be a metric space and E a subset of X. A point $P \in E$ is called an *isolated point* of E if there is an $r > 0$ such that $E \cap B(P, r) = \{P\}$. If a point of E is not isolated then it is called *non-isolated.*

We see that the notion of "isolated" has intuitive appeal: an isolated point is one that is spaced apart—by at least distance r—from the other points of the space. A non-isolated point, by contrast, has neighbors that are arbitrarily close.

EXAMPLE 9.1.14. Every point of the integers, with the usual metric, is isolated because each integer has a ball of radius $1/2$ about it that contains *only that integer.* No point of the interval [0, 1] is isolated. In the set

$$S = \left\{1, \frac{1}{2}, \frac{1}{3}, \ldots\right\} \cup \{0\},$$

every point is isolated except 0. □

Definition 9.1.15. Let (X, ρ) be a metric space and $f : X \to \mathbb{R}$. If $P \in X$ and $\ell \in \mathbb{R}$ we say that *the limit of f at P is ℓ*, and we write

$$\lim_{x \to P} f(x) = \ell,$$

if, for any $\epsilon > 0$, there is a $\delta > 0$ such that if $0 < \rho(x, P) < \delta$ then $|f(x) - \ell| < \epsilon$.

Notice in this definition that we use ρ to measure distance in X—that is the natural notion of distance with which X comes equipped—but we use absolute values to measure distance in $\mathbb{R}$.

The following lemma will prove useful.

Lemma 9.1.16. *Let (X, ρ) be a metric space and $P \in X$. Let f be a function from X to $\mathbb{R}$. Then $\lim_{x \to P} f(x) = \ell$ if and only if, for every sequence $\{x_j\} \subseteq X$ satisfying $x_j \to P$, it holds that $f(x_j) \to f(P)$.*

Definition 9.1.17. Let (X, ρ) be a metric space and E a subset of X. Suppose that $P \in E$. We say that a function $f : E \to \mathbb{R}$ is *continuous* at P if

$$\lim_{x \to P} f(x) = f(P).$$

EXAMPLE 9.1.18. Let (X, ρ) be the space of continuous functions on the interval [0, 1] equipped with the supremum metric as in Example 9.1.4 above. Define the function $\mathcal{F} : X \to \mathbb{R}$ by the formula

$$\mathcal{F}(f) = \int_0^1 f(t)\,dt.$$

Then $\mathcal{F}$ takes an element of X, namely a continuous function, to a real number, namely its integral over [0, 1]. We claim that $\mathcal{F}$ is continuous at every point of X.

Fix a point $f \in X$. If $\{f_j\}$ is a sequence of elements of X converging in the metric space sense to the limit f, then (in the language of classical analysis as in Chapter 8) the f_j are continuous functions converging uniformly to the continuous function f on the interval $[0, 1]$. But, by Theorem 8.1.8, it follows that

$$\int_0^1 f_j(t)dt \to \int_0^1 f(t)dt.$$

But this just says that $\mathcal{F}(f_j) \to \mathcal{F}(f)$. Using the lemma, we conclude that

$$\lim_{g \to f} \mathcal{F}(g) = \mathcal{F}(f).$$

Therefore $\mathcal{F}$ is continuous at f.

Since $f \in X$ was chosen arbitrarily, we conclude that the function $\mathcal{F}$ is continuous at every point of X. □

In the next section we shall develop some topological properties of metric spaces.

9.2 TOPOLOGY IN A METRIC SPACE

9.2.1 BALLS IN A METRIC SPACE

Fix a metric space (X, ρ). An *open ball* in the metric space is a set of the form

$$B(P, r) \equiv \{x \in X : \rho(x, P) < r\},$$

where $P \in X$ and $r > 0$. A set $U \subseteq X$ is called *open* if for each $u \in U$ there is an $r > 0$ such that $B(u, r) \subseteq U$.

We define a *closed ball* in the metric space (X, ρ) to be

$$\overline{B}(P, r) \equiv \{x \in X : \rho(x, P) \leq r\}.$$

A set $E \subseteq X$ is called *closed* if its complement in X is open.

EXAMPLE 9.2.1. Consider the set of real numbers $\mathbb{R}$ equipped with the metric $\rho(s, t) = 1$ if $s \neq t$ and $\rho(s, t) = 0$ otherwise. Then each singleton $U = \{x\}$ is an open set. For let P be a point of U. Then $P = x$ and the ball $B(P, 1/2)$ lies in U.

However, each singleton is also closed. For the complement of the singleton $U = \{x\}$ is the set $S = \mathbb{R} \setminus \{x\}$. If $s \in S$ then $B(s, 1/2) \subseteq S$ as in the preceding paragraph. □

EXAMPLE 9.2.2. Let (X, ρ) be the metric space of continuous functions on the interval $[0, 1]$ equipped with the metric $\rho(f, g) = \sup_{x \in [0,1]} |f(x) - g(x)|$. Define

$$U = \{f \in X : f(1/2) > 5\}.$$

Then U is an open set in the metric space. To verify this, fix an element $f \in U$. Let $\epsilon = f(1/2) - 5 > 0$. We claim that the metric ball $B(f, \epsilon)$ lies in U. Let $g \in B(f, \epsilon)$. Then

$$\begin{aligned} g(1/2) &\geq f(1/2) - |f(1/2) - g(1/2)| \\ &\geq f(1/2) - \rho(f, g) \\ &> f(1/2) - \epsilon \\ &= 5. \end{aligned}$$

It follows that $g \in U$. Since $g \in B(f, \epsilon)$ was chosen arbitrarily, we may conclude that $B(f, \epsilon) \subseteq U$. But this says that U is open.

We may also conclude from this calculation that

$$^{c}U = \{f \in X : f(1/2) \leq 5\}$$

is closed. □

EXAMPLE 9.2.3. Let X be the unit circle in the plane together with the origin 0. Equip X with the usual Euclidean metric. Then the open ball $B(0, 1)$ is just the singleton $\{0\}$. The *closure of the open ball* $B(0, 1)$ is also the singleton $\{0\}$. But the closed ball $\overline{B}(0, 1)$ is the entire space X. Thus the terminology "closed ball" is a bit confusing. □

9.2.2 ACCUMULATION POINTS

Definition 9.2.4. Let (X, ρ) be a metric space and $S \subseteq X$. A point $x \in X$ is called an *accumulation point* of S if every $B(x, r)$ contains infinitely many distinct elements of S.

Proposition 9.2.5. *Let (X, ρ) be a metric space. A set $S \subseteq X$ is closed if and only if every accumulation point of S lies in S.*

EXAMPLE 9.2.6. Let $T = [0, 1]$. Then every point of T is an accumulation point. Let

$$S = \left\{1, \frac{1}{2}, \frac{1}{3}, \dots\right\} \cup \{0\},$$

Then only the point $0 \in S$ is an accumulation point. □

Definition 9.2.7. Let (X, ρ) be a metric space. A subset $S \subseteq X$ is said to be *bounded* if S lies in some ball $B(P, r)$.

EXAMPLE 9.2.8. Consider the real numbers $\mathbb{R}$ with the usual notion of distance. Then the sets

$$\{x \in \mathbb{R} : x^3 - 3x + 7 = 0\}, \quad \{x \in \mathbb{R} : x^2 < 7\}, \quad \{x \in \mathbb{R} : 0 < x < 1/x\}$$

are all bounded. By contrast, the sets

$$\{x \in \mathbb{R} : \sin x = 0\}, \quad \{x \in \mathbb{R} : x^2 > 7\}, \quad \{x \in \mathbb{R} : x > 1/x\}$$

are all unbounded. □

9.2.3 COMPACTNESS

The definition of compact set using open coverings is universal, and we shall consider it again in what follows. But Definition 9.2.9 is the most useful characterization of compactness in a metric space.

Definition 9.2.9. Let (X, ρ) be a metric space. A set $S \subseteq X$ is said to be *compact* if every sequence in S has a subsequence that converges *to an element of* S.

EXAMPLE 9.2.10. In Chapter 4 we learned that, in the real number system, compact sets are closed and bounded, and conversely. Such is not the case in general metric spaces.

As an example, consider the metric space (X, ρ) consisting of all continuous functions on the interval [0, 1] with the supremum metric as in previous examples. Let

$$S = \{f_j(x) = x^j : j = 1, 2, \ldots\}.$$

This set is bounded since it lies in the ball $B(\mathbf{0}, 2)$ (here $\mathbf{0}$ denotes the identically zero function). We claim that S contains no Cauchy sequences. This follows (see the discussion of uniform convergence in Chapter 8) because, no matter how large N is, if $k > j > N$ then we may write

$$|f_j(x) - f_k(x)| = \left|x^j\right| \left|(x^{k-j} - 1)\right|.$$

Fix j. If x is sufficiently near to 1 then $|x^j| > 3/4$. But then we may pick k so large that $|x^{k-j}| < 1/4$. Thus

$$|f_k(x) - f_j(x)| \geq 9/16.$$

So there is no Cauchy subsequence. We may conclude (for vacuous reasons) that S is closed.

But S is not compact. For, as just noted, the sequence $\{f_j\}$ consists of infinitely many distinct elements of S which do not have a convergent subsequence (indeed not even a Cauchy subsequence). □

In spite of the last example, half of the Heine-Borel Theorem is true:

Proposition 9.2.11. *Let (X, ρ) be a metric space and S a subset of X. If S is compact then S is closed and bounded.*

Definition 9.2.12. Let S be a subset of a metric space (X, ρ). A collection of open sets $\{\mathcal{O}_\alpha\}_{\alpha \in A}$ (each $\mathcal{O}_\alpha$ is an open set in X) is called an *open covering* of S if

$$\bigcup_{\alpha \in A} \mathcal{O}_\alpha \supseteq S.$$

Definition 9.2.13. If $\mathcal{C}$ is an open covering of a set S and if $\mathcal{D}$ is another open covering of S such that each element of $\mathcal{D}$ is also an element of $\mathcal{C}$ then we call $\mathcal{D}$ a *subcovering* of $\mathcal{C}$.

We call $\mathcal{D}$ a *finite subcovering* if $\mathcal{D}$ has just finitely many elements.

Theorem 9.2.14. *A subset S of a metric space (X, ρ) is compact if and only if every open covering $\mathcal{C} = \{\mathcal{O}_\alpha\}_{\alpha \in A}$ of S has a finite subcovering.*

Proposition 9.2.15. *Let S be a compact subset of a metric space (X, ρ). If E is a closed subset of S then E is compact.*

EXAMPLE 9.2.16. Let $S = (0, 1)$. Define $U_j = \{x \in \mathbb{R} : 1/(j+3) < x < 1\}$ for $j = 1, 2, \ldots$. Then the collection $\mathcal{U} = \{U_j\}$ is an open covering of S. But there is no finite subcovering. So S is not compact. Also note that S is bounded but not closed, which gives a second reason why S is not compact. Thirdly, the sequence $s_j = 1/(j + 1)$ lies in S but has no convergent subsequence in S. Once again, S is not compact.

By contrast, the set $T = [0, 0.9]$ *is* compact. First of all, it is closed and bounded. Second, the method of bisection can be used to see that any sequence in S has a convergent subsequence. Third, any open cover of T has a finite subcover. This is tricky to prove in general, but we can look at an example:

Let $U_0 = (-0.1, 0.25)$ and let U_j for $j \geq 1$ be as in the last paragraph but one. Then $\mathcal{U} = \{U_j\}$ certainly covers T. In addition, the collection

$$\mathcal{V} = \{U_0, U_1, U_2\}$$

is a finite subcovering. □

9.3 THE BAIRE CATEGORY THEOREM

9.3.1 DENSITY

Let (X, ρ) be a metric space and $S \subseteq X$ a subset. A set $E \subseteq S$ is said to be *dense* in S if every element of S is the limit of some sequence of elements of E. Equivalently, E is dense in S if any neighborhood U of any element $s \in S$ contains points of E.

EXAMPLE 9.3.1. The set of rational numbers $\mathbb{Q}$ is dense in any open subset of the reals $\mathbb{R}$ equipped with the usual metric. □

EXAMPLE 9.3.2. Let (X, ρ) be the metric space of continuous functions on the interval $[0, 1]$ equipped with the supremum metric as usual. Let $\mathcal{P} \subseteq X$ be the polynomial functions. Then the Weierstrass Approximation Theorem tells us that $\mathcal{P}$ is dense in X. □

EXAMPLE 9.3.3. Consider the real numbers $\mathbb{R}$ with the metric $\rho(s, t) = 1$ if $s \neq t$ and $\rho(s, t) = 0$ otherwise. Then no proper subset of $\mathbb{R}$ is dense in $\mathbb{R}$. To see this, notice that if E were dense and were not all of $\mathbb{R}$ and if $P \in \mathbb{R} \setminus E$ then $\rho(P, e) > 1/2$ for all $e \in E$. So elements of E do not get close to P. Thus E is not dense in $\mathbb{R}$. □

9.3.2 CLOSURE

Definition 9.3.4. If (X, ρ) is a metric space and $E \subseteq X$ then the *closure* of E is defined to be the union of E with the set of its accumulation points.

EXAMPLE 9.3.5. Let (X, ρ) be the set of real numbers with the usual metric and set $E = \mathbb{Q} \cap (-2, 2)$. Then the closure of E is $[-2, 2]$.

Let (Y, σ) be the continuous functions on $[0, 1]$ equipped with the supremum metric as in Example 9.1.4. Take $E \subseteq Y$ to be the polynomials. Then the closure of E is the set Y. □

Definition 9.3.6. Let (X, ρ) be a metric space. We say that $E \subseteq X$ is *nowhere dense* in X if the closure of E contains no ball $B(x, r)$ for any $x \in X, r > 0$.

EXAMPLE 9.3.7. Let us consider the integers $\mathbb{Z}$ as a subset of the metric space $\mathbb{R}$ equipped with the standard metric. Then the closure of $\mathbb{Z}$ is $\mathbb{Z}$ itself. And of course $\mathbb{Z}$ contains no metric balls. Therefore $\mathbb{Z}$ is nowhere dense in $\mathbb{R}$. □

EXAMPLE 9.3.8. Consider the metric space X of all continuous functions on the unit interval $[0, 1]$, equipped with the usual supremum metric. Fix $k > 0$ and consider

$$E_k \equiv \{p(x) : p \text{ is a polynomial of degree not exceeding } k\}.$$

Then the closure of E_k is E_k itself (that is, the limit of a sequence of polynomials of degree not exceeding k is still a polynomial of degree not exceeding k). And E_k contains no metric balls. For if $p \in E_k$ and $r > 0$ then $p(x) + (r/2) \cdot x^{k+1} \in B(p, r)$ but $p(x) + (r/2) \cdot x^{k+1} \notin E$.

We recall, as noted in Example 9.3.2, that the set of *all* polynomials is dense in X; but if we restrict attention to polynomials of degree not exceeding a fixed integer k then the resulting set is nowhere dense. □

9.3.3 BAIRE'S THEOREM

Theorem 9.3.9. *Let (X, ρ) be a complete metric space. Then X cannot be written as the union of countably many nowhere dense sets.*

Before we apply the Baire Category Theorem, let us formulate some restatements, or corollaries, of the theorem that follow immediately from the definitions.

Corollary 9.3.10. *Let (X, ρ) be a complete metric space. Let $Y_1, Y_2, \ldots$ be countably many closed subsets of X each of which contains no non-trivial open ball. Then $\bigcup_j Y_j$ also contains no non-trivial open ball.*

Corollary 9.3.11. *Let (X, ρ) be a complete metric space. Let $O_1, O_2, \ldots$ be countably many dense open subsets of X. Then $\bigcap_j O_j$ is dense in X.*

The result of the second corollary follows from the first corollary by complementation. The set $\bigcap_j O_j$, while dense, need not be open.

EXAMPLE 9.3.12. The metric space $\mathbb{R}$, equipped with the standard Euclidean metric, cannot be written as a countable union of nowhere dense sets. □

By contrast, $\mathbb{Q}$ *can* be written as the union of the singletons $\{q_j\}$ where the q_j represent an enumeration of the rationals. And of course each $\{q_j\}$ is nowhere dense. However, $\mathbb{Q}$ is not complete.

EXAMPLE 9.3.13. Baire's theorem contains the fact that a perfect set of real numbers must be uncountable. For if P is perfect and countable we may write $P = \{p_1, p_2, \ldots\}$. Therefore

$$P = \bigcup_{j=1}^{\infty} \{p_j\}.$$

But each of the singletons $\{p_j\}$ is a nowhere dense set in the metric space P. And P is complete. (You should verify both these assertions for yourself.) This contradicts the category theorem. So P cannot be countable. □

A set that can be written as a countable union of nowhere dense sets is said to be of *first category*. If a set is not of first category, then it is said to be of *second category*. The Baire Category Theorem says that a complete metric space must be of second category. We should think of a set of first category as being "thin" and a set of second category as being "fat" or "robust." (This is one of many ways that we have in mathematics of distinguishing "fat" sets. Countability and uncountability is another. Lebesgue's measure theory, not covered in this book, is a third.)

One of the most striking applications of the Baire Category Theorem is the following result to the effect that "most" continuous functions are nowhere differentiable. This explodes the myth that most of us learn in calculus that a typical function is differentiable at all points except perhaps at a discrete set of bad points.

Theorem 9.3.14. *Let* (X, ρ) *be the metric space of continuous functions on the unit interval* $[0, 1]$ *equipped with the metric*

$$\rho(f, g) = \sup_{x \in [0,1]} |f(x) - g(x)|.$$

Define a subset of E *of* X *as follows:* $f \in E$ *if there exists one point at which* f *is differentiable. Then* E *is of first category in the complete metric space* (X, ρ).

9.4 THE ASCOLI-ARZELA THEOREM

9.4.1 EQUICONTINUITY

Let $\mathcal{F} = \{f_\alpha\}_{\alpha \in A}$ be a family, not necessarily countable, of functions on a metric space (X, ρ). We say that the family $\mathcal{F}$ is *equicontinuous* on X if, for every $\epsilon > 0$, there is a $\delta > 0$ such that, when $\rho(s, t) < \delta$, then $|f_\alpha(s) - f_\alpha(t)| < \epsilon$. Equicontinuity mandates not only uniform continuity of each f_α but also that the uniformity occurs simultaneously, and at the same rate, for all the f_α.

EXAMPLE 9.4.1. Let (X, ρ) be the unit interval $[0, 1]$ with the usual Euclidean metric. Let $\mathcal{F}$ consist of all functions f on X that satisfy the Lipschitz condition

$$|f(s) - f(t)| \le 2 \cdot |s - t|$$

for all s, t. Then $\mathcal{F}$ is an equicontinuous family of functions. For if $\epsilon > 0$ then we may take $\delta = \epsilon/2$. Then if $|s - t| < \delta$ and $f \in \mathcal{F}$ we have

$$|f(s) - f(t)| \le 2 \cdot |s - t| < 2 \cdot \delta = \epsilon.$$

The mean value theorem tells us that $\sin x, \cos x, 2x, x^2$ are elements of $\mathcal{F}$. □

9.4.2 EQUIBOUNDEDNESS

If $\mathcal{F}$ is a family of functions on X we call $\mathcal{F}$ *equibounded* if there is a number $M > 0$ such that

$$|f(x)| \le M$$

for all $x \in X$ and all $f \in \mathcal{F}$. For example, the functions $f_j(x) = \sin jx$ on $[0, 1]$ form an equibounded family.

9.4.3 THE ASCOLI-ARZELA THEOREM

One of the cornerstones of classical analysis is the following result of Ascoli and Arzela:

Theorem 9.4.2. *Let (Y, σ) be a compact metric space. Let $\mathcal{F}$ be an equibounded, equicontinuous family of functions on Y. Then there is a sequence $\{f_j\} \subseteq \mathcal{F}$ that converges uniformly to a continuous function on Y.*

Let (X, ρ) be the metric space consisting of the continuous functions on the unit interval $[0, 1]$ equipped with the usual supremum norm. Let $\mathcal{F}$ be a closed, equicontinuous, equibounded family of functions lying in X. Then the theorem says that $\mathcal{F}$ is a compact set in this metric space. For any infinite subset of $\mathcal{F}$ is guaranteed to have a convergent subsequence with limit in $\mathcal{F}$. As a result, we may interpret the Ascoli-Arzela theorem as identifying certain compact collections of continuous functions.

EXAMPLE 9.4.3. Refer, for instance, to Example 9.4.1. The set $\mathcal{F}$ of functions on $[0, 1]$ that are bounded by 2 and satisfy the Lipschitz condition

$$|f(s) - f(t)| \le 2|s - t|$$

forms an equibounded, equicontinuous family in the metric space (X, ρ) of continuous functions on the unit interval with the usual uniform metric. By the Ascoli-Arzela theorem, every sequence in $\mathcal{F}$ has a convergent subsequence. $\square$

It is common in the theory of partial differential equations to derive the existence of a solution by first proving an *a priori* estimate for smooth functions and then extracting a solution in general, using the Ascoli-Arzela theorem, as the limit of smooth solutions.

Glossary of Terms from Real Variable Theory

absolutely convergent: A series $\sum_j c_j$ is absolutely convergent if $\sum_j |c_j|$ converges.

absolute maximum: If f is a function with domain S and if there is a point $M \in S$ such that $f(M) \geq f(x)$ for all $x \in S$ then the point M is called an absolute maximum for f.

absolute minimum: If f is a function with domain S and if there is a point $m \in S$ such that $f(m) \leq f(x)$ for all $x \in S$ then the point m is called an absolute minimum for f.

accumulation point: Let S be a set. A point x is called an accumulation point of S if every neighborhood of x contains infinitely many distinct elements of S.

accumulation point in a metric space: Let (X, ρ) be a metric space and S a subset. A point $x \in X$ is called an accumulation point of S if every $B(x, r)$ contains infinitely many elements of S.

bijection: A one-to-one, onto mapping.

boundary point: Let S be a set. Then B is a boundary point of S if every non-empty neighborhood $(b-\epsilon, b+\epsilon)$ contains both points of S and points of $\mathbb{R} \setminus S$.

bounded above: A set $S \subseteq \mathbb{R}$ is called bounded above if there is a number M such that $s \leq M$ for every element $s \in S$.

bounded below: A set $S \subseteq \mathbb{R}$ is called bounded below if there is a number N such that $s \geq N$ for every element $s \in S$.

bounded set: A set $S \subseteq \mathbb{R}$ is called bounded if there is a positive number K such that $|s| \leq K$ for every element $s \in S$.

bounded set in a metric space: Let (X, ρ) be a metric space. A subset $S \subseteq X$ is said to be bounded if S lies in some ball $B(P, r)$.

bounded variation: Let f be a function on the interval $[a, b]$. For $x \in [a, b]$ we define $Vf(x) = \sup \sum_{j=1}^{k} \left| f(p_j) - f(p_{j-1}) \right|$, where the supremum is taken over all partitions $\mathcal{P}$ of the interval $[a, x]$. If $Vf \equiv Vf(b) < \infty$, then the function f is said to be of bounded variation on the interval $[a, b]$.

cardinality: A measure (due to Cantor) of the size of a set.

cartesian product: The collection of ordered pairs, or n-tuples, of objects from given sets.

Cauchy criterion: We say that the sequence $\{a_j\}$ satisfies the Cauchy criterion if, for each $\epsilon > 0$, there is an $N > 0$ such that, whenever $j, k > N$, then $|a_j - a_k| < \epsilon$.

Cauchy criterion in a metric space: A sequence $\{x_j\}$ of elements of a metric space (X, ρ) is said to be Cauchy if, for each $\epsilon > 0$, there is an $N > 0$ such that, if $j, k > N$, then $\rho(x_j, x_k) < \epsilon$.

closed interval: A set of the form $[a, b] = \{x \in \mathbb{R} : a \leq x \leq b\}$.

closed set: A set whose complement is open.

closure of a set in a metric space: Let (X, ρ) be a metric space and $E \subseteq X$. The closure of E is defined to be the union of E with the set of its accumulation points. We sometimes denote the closure of E by $\overline{E}$.

common refinement: If $\mathcal{P}$, $\mathcal{P}'$ are partitions of $[a, b]$ then their common refinement is the union of all the points of $\mathcal{P}$ and $\mathcal{P}'$.

commuting limits: For example, if

$$\lim_{x \to c} \lim_{j \to \infty} y_j(x) = \lim_{j \to \infty} \lim_{x \to c} y_j(x),$$

then the x and j limits are said to commute.

compact set: A set S is compact if every sequence in S has a subsequence that converges to an element of S.

compact set in a metric space: Let (X, ρ) be a metric space. A set $S \subseteq X$ is said to be compact if every sequence in S has a subsequence that converges to an element of S.

complement of a set: The collection of elements in the universe under consideration that are not in that set.

complete metric space: A metric space (X, ρ) is called complete if every Cauchy sequence in X converges to an element of the metric space.

conditionally convergent: A series $\sum_j c_j$ is conditionally convergent if $\sum_j c_j$ converges but $\sum_j |c_j|$ diverges.

connected: If a set is not disconnected then it is connected.

continuous A function f is continuous at P if $\lim_{x \to P} f(x) = f(P)$.

convergence of a sequence: A sequence $\{a_j\}$ converges to a limit ℓ if, for every $\epsilon > 0$, there is a positive integer N such that $j > N$ implies that $|a_j - \ell| < \epsilon$.

convergence of a sequence in a metric space: A sequence $\{x_j\}$ of elements of the metric space (X, ρ) is said to converge to a point $\alpha \in X$ if, for each $\epsilon > 0$, there is an $N > 0$ such that if $j > N$ then $\rho(x_j, \alpha) < \epsilon$.

convergence of a series: A series $\sum_j c_j$ converges if the sequence of partial sums $\{S_N\}$, with $S_N = \sum_{j=1}^{N} c_j$, converges to a finite limit.

convergence of a series of functions: In case the limit of partial sums $\lim_{N\to\infty} S_N(x)$ exists and is finite we say that the series $\sum_j f_j(x)$ converges at x.

converge pointwise: A sequence of functions converges pointwise to a limit function f on S if, for each $x \in S$, the sequence of numbers $\{f_j(x)\}$ converges to $f(x)$.

converge uniformly: A sequence of functions $\{f_j\}$ converges uniformly to f if, given $\epsilon > 0$, there is an $N > 0$ such that for any $j > N$ and any $x \in S$ it holds that $|f_j(x) - f(x)| < \epsilon$.

cosine function: The power series function $\sum_{j=0}^{\infty} (-1)^j \frac{x^{2j}}{(2j)!}$.

countable: A set that has the same cardinality as the set of natural numbers.

derivative: If f is a function with domain an open interval I and if $x \in I$ then the limit $\lim_{t\to x} \frac{f(t)-f(x)}{t-x}$, when it exists, is called the derivative of f at x.

difference of two functions: If f, g are given functions, then $f - g$ is their difference.

differentiable: If the derivative of f at x exists then we say that f is differentiable at x.

disconnected: A set S is disconnected if it is possible to find a pair of nonempty open sets U and V such that $U \cap S \neq \emptyset$, $V \cap S \neq \emptyset$, $(U \cap S) \cap (V \cap S) = \emptyset$, and $S = (U \cap S) \cup (V \cap S)$.

discontinuity of the first kind: If $\lim_{x \to P^-} f(x)$ and if $\lim_{x \to P^+} f(x)$ exist but either do not equal each other or do not equal $f(P)$ then we say that f has a discontinuity of the first kind.

discontinuity of the second kind: If either $\lim_{x \to P^-}$ does not exist or $\lim_{x \to P^+}$ does not exist then we say that f has a discontinuity of the second kind at P.

divergence of a sequence: A sequence that does not converge instead diverges.

divergence of a series: If the sequence of partial sums of a series does not converge then it diverges.

divergence of a series of functions: If the sequence of partial sums of a series of functions does not converge then it diverges.

domain of a function: The set on which a function operates.

element of a set: An object in that set.

empty set: The set with no elements.

equibounded: A family $\{f_\alpha\}_{\alpha \in A}$ of real-valued functions on a metric space (X, ρ) is called *equibounded* if there is an number $M > 0$ so that $|f_\alpha(x)| \leq M$ for every $x \in X$ and every $\alpha \in A$.

equicontinuous: A family $\{f_\alpha\}_{\alpha \in A}$ of real-valued functions on a metric space (X, ρ) is called *equicontinuous* if, for any $\epsilon > 0$, there is $\delta > 0$ so that if $x, y \in X$ with $\rho(x, y) < \delta$ and $\alpha \in A$ then $|f_\alpha(x) - f_\alpha(y)| < \epsilon$.

exponential function: The power series function $\exp(z) = \sum_{j=0}^{\infty} \frac{z^j}{j!}$.

field: A set with operations of addition and multiplication, satisfying the usual laws of arithmetic.

finite subcovering: A subcovering with finitely many elements.

finite subcovering in a metric space: Let S be a set in a metric space (X, ρ) and let $\mathcal{C}$ be an open covering of S. We call $\mathcal{D}$ a finite subcovering of $\mathcal{C}$ if $\mathcal{D}$ is a subcovering and if $\mathcal{D}$ has just finitely many elements.

first category: A set is of first category if it can be written as the countable union of nowhere dense sets.

function: A rule that assigns objects from one set to the elements of another.

geometric series: A series of the form $\sum_j \alpha^j$.

greatest integer function: The function $\lfloor x \rfloor$ that equals the greatest integer that does not exceed x.

greatest lower bound: A lower bound for a set such that there is no other lower bound greater than it.

Hadamard formula: For the power series $\sum_{j=0}^{\infty} a_j (x-c)^j$, Hadamard defines A and ρ by

$$A = \limsup_{n\to\infty} |a_n|^{1/n},$$

$$\rho = \begin{cases} 0 & \text{if } A = \infty, \\ 1/A & \text{if } 0 < A < \infty, \\ \infty & \text{if } A = 0. \end{cases}$$

harmonic series: The series $\sum_j 1/j$.

image If f is a function and G a set then the image $f(G)$ is the set of $f(x)$ such that $x \in G$.

infimum: See greatest lower bound.

integers: The positive and negative whole numbers and zero.

interior point: Let S be a set. A point s is called an interior point of S if there is an $\epsilon > 0$ such that the interval $(s-\epsilon, s+\epsilon)$ lies in S.

intersection: Those elements common to two or more given sets.

interval of convergence: The interval centered at c on which a power series expanded about c converges.

inverse function: Given a function f, the inverse sends range elements of f to their corresponding domain elements.

inverse image If f is a function and $\mathcal{O}$ a set then the inverse image $f^{-1}(\mathcal{O})$ is the set of x such that $f(x) \in \mathcal{O}$.

isolated point Let S be a set. A point t is called an isolated point of S if there is an $\epsilon > 0$ such that the intersection of the interval $(t-\epsilon, t+\epsilon)$ with S is just the singleton $\{t\}$.

isolated point in a metric space: A point P of a set E in a metric space (X, ρ) is called isolated if there is an $r > 0$ such that $E \cap B(P, r) = \{P\}$.

***k* times continuously differentiable:** If f is k times differentiable on an open interval I and if each of the derivatives $f^{(1)}, f^{(2)}, \ldots, f^{(k)}$ is continuous on I then we say that f is k times continuously differentiable on I.

least upper bound: An upper bound for a set such that there is no other upper bound less than it.

left limit: The function f on E has left limit ℓ at P, and we write

$$\lim_{x \to P^-} f(x) = \ell,$$

if for every $\epsilon > 0$ there is a $\delta > 0$ such that whenever $P - \delta < x < P$ and $x \in E$ then it holds that $|f(x) - \ell| < \epsilon$.

limit of a function: The function f with domain $E \subseteq \mathbb{R}$ has limit ℓ at P if, for each $\epsilon > 0$, there is a $\delta > 0$ such that when $x \in E$ and $0 < |x - P| < \delta$ then $|f(x) - \ell| < \epsilon$.

limit of a function on a metric space: Let (X, ρ) be a metric space and f a function on it. If, for any $\epsilon > 0$, there is a $\delta > 0$ such that, if $0 < \rho(x, P) < \delta$, then $|f(x) - \ell| < \epsilon$, then we say that f has limit ℓ at P.

limit of a sequence: If the sequence $\{a_j\}$ converges to ℓ then we call ℓ the limit of the sequence.

limit of a sequence in a metric space: If the sequence $\{x_j\}$ of elements of a metric space converges to α then we call α the limit of the sequence.

limit infimum: The least limit of all subsequences of the given sequence.

limit supremum: The greatest limit of all subsequences of the given sequence.

Lipschitz condition of order α: Let f be a function on the interval I. There is a constant M such that for all $s, t \in I$ we have $|f(s) - f(t)| \leq M \cdot |s - t|^{\alpha}$. Here $0 < \alpha \leq 1$.

local extrema: Local maxima and local minima.

local maximum: A point $x \in (a, b)$ is called a local maximum for f if there is a $\delta > 0$ such that $f(t) \leq f(x)$ for all $t \in (x - \delta, x + \delta)$.

local minimum: A point $x \in (a, b)$ is called a local minimum for f if there is a $\delta > 0$ such that $f(t) \geq f(x)$ for all $t \in (x - \delta, x + \delta)$.

lower bound: A number that is less than or equal to all elements of a given set.

lower Riemann integral: Let f be a function on the interval $[a, b]$. Define the lower Riemann integral $I^*(f) = \sup \mathcal{L}(f, \mathcal{P}, \alpha)$, where the supremum is taken with respect to all partitions of the interval $[a, b]$.

lower Riemann sum: Fix an interval $[a, b]$ and a monotonically increasing function α on $[a, b]$. If $\mathcal{P} = \{p_0, p_1, \ldots, p_k\}$ is a partition of $[a, b]$, let $\Delta\alpha_j = \alpha(p_j) - \alpha(p_{j-1})$. Let f be a bounded function on $[a, b]$ and define the lower Riemann sum of f with respect to α as follows: $\mathcal{L}(f, \mathcal{P}, \alpha) = \sum_{j=1}^{k} m_j \Delta\alpha_j$. Here m_j denotes the infimum of f on I_j.

mesh: If $\mathcal{P} = \{x_0, x_1, \ldots, x_k\}$ is a partition of $[a, b]$ we let I_j denote the interval $[x_{j-1}, x_j]$, $j = 1, 2, \ldots, k$. The symbol Δ_j denotes the length of I_j. The mesh of $\mathcal{P}$, denoted by $m(\mathcal{P})$, is defined to be $\max \Delta_j$.

metric: The function ρ in the definition of metric space.

metric space: A metric space is a pair (X, ρ), where X is a set and $\rho : X \times X \to \{t \in \mathbb{R} : t \geq 0\}$ is a function satisfying

1. $\forall x, y \in X, \rho(x, y) = \rho(y, x)$;
2. $\rho(x, y) = 0$ if and only if $x = y$;
3. $\forall x, y, z \in X, \rho(x, y) \leq \rho(x, z) + \rho(z, y)$.

monotonically decreasing function: The function f is monotonically decreasing on (a, b) if, whenever $a < s < t < b$, it holds that $f(s) \geq f(t)$.

monotone decreasing sequence: The sequence $\{a_j\}$ is monotone decreasing if $a_1 \geq a_2 \geq \ldots$.

monotonically increasing function: The function f is monotonically increasing on (a, b) if, whenever $a < s < t < b$, it holds that $f(s) \leq f(t)$.

monotonically increasing sequence: The sequence $\{a_j\}$ is monotone increasing if $a_1 \leq a_2 \leq \ldots$.

natural logarithm function: The inverse of the exponential function.

natural numbers: The whole, or counting, numbers.

neighborhood: If $x \in \mathbb{R}$, then a neighborhood of x is an open set containing x.

non-isolated point: A point that is not isolated.

non-isolated point in a metric space: A point (in a metric space) that is not isolated.

nowhere dense set in a metric space: Let (X, ρ) be a metric space. The set $E \subseteq X$ is nowhere dense in X if the closure of E contains no ball $B(x, r)$ for any $x \in X, r > 0$.

one-to-one: A function that sends different domain values to different range values. In other words, if $f(a) = f(b)$, then $a = b$.

onto: A function that assumes all values in its range.

open covering: A collection of open sets $\{\mathcal{O}_\alpha\}_{\alpha \in A}$ is called an open covering of S if $\bigcup_{\alpha \in A} \mathcal{O}_\alpha \supseteq S$.

open covering in a metric space: Let S be a subset of a metric space (X, ρ). A collection of open sets $\{\mathcal{O}_\alpha\}_{\alpha \in A}$ (each $\mathcal{O}_\alpha$ is an open set in X) is called an open covering of S if $\cup_{\alpha \in A} \mathcal{O}_\alpha \supseteq S$.

open interval: A set of the form $(a, b) = \{x \in \mathbb{R} : a < x < b\}$.

open set: A set with the property that, whenever $x \in U$, there is an $\epsilon > 0$ such that $(x - \epsilon, x + \epsilon) \subseteq U$.

ordered field: A field equipped with an order relation that is compatible with the field operations.

partial sum: The sum of finitely many terms of a series.

partial sum for a series of functions: The expression

$$S_N(x) = \sum_{j=1}^{N} f_j(x) = y_1(x) + y_2(x) + \cdots + y_N(x)$$

is called the Nth partial sum for the series $\sum_j f_j(x)$.

partition: Let $[a, b]$ be a closed interval in $\mathbb{R}$. A finite, ordered set of points $\mathcal{P} = \{x_0, x_1, x_2, \ldots, x_{k-1}, x_k\}$ such that $a = x_0 \leq x_1 \leq x_2 \leq \cdots \leq x_{k-1} \leq x_k = b$ is called a partition of $[a, b]$.

perfect set: A set S is perfect if it is non-empty, closed, and if every point of S is an accumulation point of S.

product of two functions: If f, g are given functions, then $f \cdot g$ is their product.

power sequence: A sequence $\{\lambda^j\}$ of powers.

power series expanded about the point c**:** A series of the form

$$\sum_{j=0}^{\infty} a_j (x-c)^j$$

is called a power series expanded about the point c.

radius of convergence: Half the length of the interval of convergence.

range of a function: The set in which a function takes its values.

rational function A quotient of polynomials.

rational numbers: The collection of quotients of integers (with division by zero disallowed).

real analytic function: A function f, with domain an open set $U \subseteq \mathbb{R}$ and range either the real or the complex numbers, such that for each $c \in U$ the function f may be represented by a convergent power series on an interval of positive radius centered at c : we have $f(x) = \sum_{j=0}^{\infty} a_j (x-c)^j$.

real power: The exponential a^x with real base a and arbitrary exponent x.

rearrangement of a series: The same sum, with the terms in a different order.

refinement: Let $\mathcal{P}$ and $\mathcal{Q}$ be partitions of the interval $[a, b]$. If each point of $\mathcal{P}$ is also an element of $\mathcal{Q}$ then we call $\mathcal{Q}$ a refinement of $\mathcal{P}$.

Riemann integrable: A function f is Riemann integrable on $[a, b]$ if the Riemann sums of $\mathcal{R}(f, \mathcal{P})$ tend to a limit as the mesh of $\mathcal{P}$ tends to zero.

Riemann integral: The value of the limit of the Riemann sums, when that limit exists.

Riemann-Stieltjes integral: When the upper and lower Riemann integrals are equal, we denote this quantity by $\int_a^b f \, d\alpha$ and call it the Riemann-Stieltjes integral.

Riemann sum If f is a function on $[a, b]$ and $\mathcal{P}$ a partition with increment lengths Δ_j then the corresponding Riemann sum is defined to be $\mathcal{R}(f, \mathcal{P}) = \sum_{j=1}^{k} f(s_j)\Delta_j$ for points s_j in the intervals I_j of the partition.

right limit: The function f on E has right limit ℓ at P, and we write $\lim_{x \to P^+} f(x) = \ell$, if for every $\epsilon > 0$ there is a $\delta > 0$ such that whenever $P < x < P + \delta$ and $x \in E$ then it holds that $|f(x) - \ell| < \epsilon$.

same cardinality: Two sets with a bijection between them.

second category: A set is of second category if it is not of first category.

sequence on a set S: An ordered list of numbers, or a function f from $\mathbb{N}$ to S.

sequence of functions: A sequence whose terms are functions, usually written $f_1(x), f_2(x), \ldots$ or $\{f_j\}_{j=1}^{\infty}$.

series: An infinite sum.

series of functions: The formal expression $\sum_{j=1}^{\infty} y_j(x)$, where the y_j are functions on a common domain S, is called a series of functions.

set-builder notation: Specification of a set with the notation $S = \{x \in \mathbb{R} : P(x)\}$, where P is a property that the number x may or may not have.

set-theoretic difference: Given two sets, the collection of objects in one set but not in the other.

set: A collection of objects.

simple discontinuity: See **discontinuity of the first kind**.

sine function: The power series function $\sum_{j=0}^{\infty}(-1)^j \frac{x^{2j+1}}{(2j+1)!}$.

singleton: A set with one element.

subcovering: If $\mathcal{C}$ is an open covering of a set S and if $\mathcal{D}$ is another open covering of S such that each element of $\mathcal{D}$ is also an element of $\mathcal{C}$ then we call $\mathcal{D}$ a subcovering of $\mathcal{C}$.

subcovering in a metric space: Let (X, ρ) be a metric space. If $\mathcal{C}$ is an open covering of a set $S \subseteq X$ and if $\mathcal{D}$ is another open covering of S such that each element of $\mathcal{D}$ is also an element of $\mathcal{C}$ then we call $\mathcal{D}$ a subcovering of $\mathcal{C}$.

subsequence: Let $\{a_j\}$ be a given sequence. If $0 < j_1 < j_2 < \ldots$ are positive integers then the function $k \mapsto a_{j_k}$ is called a subsequence of the given sequence.

subset: A subcollection of objects in a given set.

summation by parts: A summation procedure that is analogous to integration by parts.

sum of two functions: If f, g are given functions, then $f + g$ is their sum.

supremum: See **least upper bound:**.

totally disconnected: A set S is totally disconnected if, for each distinct $x \in S$, $y \in S$, there exist disjoint, nonempty, open sets U and V such that $x \in U, y \in V$, and $S = (U \cap S) \cup (V \cap S)$.

total variation: The quantity $Vf(b)$ in the definition of bounded variation.

triangle inequality: On a metric space with metric ρ, the inequality $\rho(x, y) \leq \rho(x, z) + \rho(z, y)$.

uncountable: An infinite set with cardinality at least as great as $\mathbb{R}$.

uniform convergence of a sequence of functions: A sequence of functions f_j converges uniformly to f if, given $\epsilon > 0$, there is an $N > 0$ such that for any $j > N$ and any $x \in S$ it holds that $|f_j(x) - f(x)| < \epsilon$.

uniform convergence of a series of functions: If the partial sums $S_N(x)$ of the series $\sum_j f_j(x)$ converge uniformly on S to a limit function $g(x)$ then we say that the series converges uniformly on S.

uniformly Cauchy: A sequence of functions f_j on a domain S is called uniformly Cauchy if, for each $\epsilon > 0$, there is an $N > 0$ such that, if $j, k > N$, then $\left|f_j(x) - f_k(x)\right| < \epsilon$ for all $x \in S$.

uniformly continuous: A function f is uniformly continuous on a set E if, for any $\epsilon > 0$, there is a $\delta > 0$ such that whenever $s, t \in E$ and $|s - t| < \delta$ then $|f(s) - f(t)| < \epsilon$.

union: Those elements in any one of a collection of given sets.

upper bound: A number that exceeds all elements of a given set.

upper Riemann integral: Let f be a function on the interval $[a, b]$. Define $I^*(f) = \inf \mathcal{U}(f, \mathcal{P}, \alpha)$, where the infimum is taken with respect to all partitions of the interval $[a, b]$.

upper Riemann sum: Fix an interval $[a, b]$ and a monotonically increasing function α on $[a, b]$. If $\mathcal{P} = \{p_0, p_1, \ldots, p_k\}$ is a partition of $[a, b]$, let $\Delta\alpha_j = \alpha(p_j) - \alpha(p_{j-1})$. Let f be a bounded function on $[a, b]$ and define the upper Riemann sum of f with respect to α as follows: $\mathcal{U}(f, \mathcal{P}, \alpha) = \sum_{j=1}^{k} M_j \Delta\alpha_j$. Here M_j denotes the supremum of f on I_j.

Venn diagram: A figure that displays sets as regions in the plane.

BIBLIOGRAPHY

[ABR] R. Abraham and J. Robbin, *Transversal Mappings and Flows*, Benjamin, New York, 1967.

[BOA] R. P. Boas. *A Primer of Real Functions.* Carus Mathematical Monograph No. 13, John Wiley and Sons, Inc., New York, 1960.

[BUC] R. C. Buck. *Advanced Calculus.* 2d ed., McGraw-Hill Book Company, New York, 1965.

[BUB] P. Butzer and H. Berens, *Semi-Groups of Operators and Approximation*, Springer-Verlag, Berlin and New York, 1967.

[HOF] K. Hoffman. *Analysis in Euclidean Space.* Prentice-Hall, Inc., Englewood Cliffs, N.J., 1962.

[KRA1] S. G. Krantz, *Real Analysis and Foundations*, CRC Press, Boca Raton, Florida, 1991.

[KRA2] S. G. Krantz, *Handbook of Logic and Proof Techniques for Computer Science*, Birkhäuser, Boston, 2002.

[KRA3] S. G. Krantz, Lipschitz spaces, smoothness of functions, and approximation theory, *Expositiones Math.* 3(1983), 193–260.

[KRP] S. G. Krantz and H. R. Parks, *A Primer of Real Analytic Functions*, Birkhäuser, Boston, 2002.

[NIV] I. Niven. *Irrational Numbers.* Carus Mathematical Monograph No. 11, John Wiley and Sons, Inc., New York, 1956.

[RUD] W. Rudin. *Principles of Mathematical Analysis.* 3d ed., McGraw-Hill Book Company, New York, 1976.

[STR] K. Stromberg, *An Introduction to Classical Real Analysis.* Wadsworth Publishing, Inc., Belmont, Ca., 1981.

[SIM] G. F. Simmons and S. G. Krantz, *Differential Equations: Theory, Technique, and Practice*, McGraw-Hill, New York, 2006.

Index

About the Author

Steven G. Krantz was born in San Francisco, California in 1951. He received the B.A. degree from the University of California at Santa Cruz in 1971 and the Ph.D. from Princeton University in 1974.

Krantz has taught at UCLA, Penn State, Princeton University, and Washington University in St. Louis. He served as Chair of the latter department for five years.

Krantz has published more than 50 books and more than 150 scholarly papers. He is the recipient of the Chauvenet Prize and the Beckenbach Book Award of the MAA. He has received the UCLA Alumni Foundation Distinguished Teaching Award and the Kemper Award. He has directed 17 Ph.D. theses and 9 Masters theses.